HIGHWAY SCENE SAFETY

HIGHWAY SCENE SAFETY

Leslie J. Greenwood

DELMAR
CENGAGE Learning™

Australia • Brazil • Japan • Korea • Mexico • Singapore • Spain • United Kingdom • United States

DELMAR
CENGAGE Learning™

Highway Scene Safety
Leslie J. Greenwood

Vice President, Career and
Professional Editorial:
Dave Garza

Director of Learning Solutions:
Sandy Clark

Senior Acquisitions Editor:
Janet Maker

Managing Editor: Larry Main

Senior Product Manager:
Jennifer A. Starr

Editorial Assistant: Amy Wetsel

Vice President, Career and
Professional Marketing:
Jennifer Baker

Marketing Director:
Deborah S. Yarnell

Senior Marketing Manager:
Erin Coffin

Associate Marketing Manager:
Shanna Gibbs

Production Director:
Wendy Troeger

Production Manager:
Mark Bernard

Content Project Management:
PreMediaGlobal

Art Director:
Benjamin Gleeksman

For product information and
technology assistance, contact us at **Cengage Learning
Customer & Sales Support, 1-800-354-9706**

For permission to use material from this text or product,
submit all requests online at **cengage.com/permissions**
Further permissions questions can be emailed to
permissionrequest@cengage.com

Library of Congress Control Number: 2010938334

ISBN-13: 978-1-4354-6976-1

ISBN-10: 1-4354-6976-3

Delmar
5 Maxwell Drive
Clifton Park, NY 12065-2919
USA

Cengage Learning is a leading provider of customized learning solutions with office locations around the globe, including Singapore, the United Kingdom, Australia, Mexico, Brazil, and Japan. Locate your local office at: **international.cengage.com/region**

Cengage Learning products are represented in Canada by Nelson Education, Ltd.

To learn more about Delmar, visit **www.cengage.com/delmar**

Purchase any of our products at your local college store or at our preferred online store **www.cengagebrain.com**

Printed in the United States of America
1 2 3 4 5 6 7 15 14 13 12 11

Contents

CHAPTER 1
The Importance of Safety at Highway Incidents

CHAPTER 2
Traffic Management

CHAPTER 3
Safety Officer Considerations

CHAPTER 4
Flagging Operations

CHAPTER 5
Incident Organization

CHAPTER 6
Apparatus Considerations   89

CHAPTER 7
Preplanning Considerations

Preface

Intent of this Book

Highway Scene Safety was written with the thought in mind that all first responders will be able to work together to create a safe environment on our highways. As such, this book focuses on effective highway management while on the scene of an incident. It is intended to give our responders the best information possible and is a valuable learning tool for emergency response safety courses or a handy reference manual for the field.

In addition, *Highway Scene Safety* is designed to follow the 2009 Edition of the *Manual on Uniform Traffic Control Devices* (MUTCD), the national unified goals and other national standards. This book provides new responders with the necessary tools, and gives seasoned responders the essential edge they need to keep our highway scenes safe.

As first responders, you are charged with making emergency scenes safe, not only for the citizens you serve, but for all responders involved. The use of the skills and knowledge in this publication will help you make a difference in everyone's life.

Why I Wrote This Book

Highway Scene Safety was written to expand the knowledge of emergency responders and bring all services together in order to achieve the same operational goals. This book was designed to help responders meet standards such as the National Fire Protection Association (NFPA) 1001 and 1500 Standards and the MUTCD.

It was also designed with the intent to help reduce the line of duty injuries and deaths for all emergency responders, and covers preplanning, the national incident command system, governing laws, interagency communications and other topics that are requirements for a safe response.

How This Book is Organized

Highway Scene Safety has been organized to include all the essential information that is needed by first responders. Critical topics are addressed in a logical progression, including the following:

- The importance of highway safety
- How to manage traffic
- The role of the safety officer on the scene of an incident
- Flagging operations
- How to implement the incident command system into your operations
- How apparatus design can affect your safety
- Preplanning considerations

Features of This Book

This book includes several features that emphasize safety on the scene of an incident:

- *Case Studies* based on NIOSH reports and author experience open each chapter and highlight important lessons learned
- *Focuses on 16 Firefighter Life Safety Initiatives* created by the National Fallen Firefighters Foundation to reduce line of duty injuries and deaths
- *Highlights the National Incident Command System* and stresses the importance of working and planning with other agencies.
- *A Chapter on Flagging Operations* explains how, when and where flagging operations should be conducted to keep both responders and citizen safe
- *Current Information* regarding laws related to highway incidents, including D.O.T., vehicle and traffic, and the Manual on Unified Traffic Control Devices (MUTCD) laws.
- *Addresses the Challenges* of conducting operations on the highway, such as helicopter landings, maintaining needs of stopped traffic, placement of signage.

- *Provides Practical Guidelines* for outlining the traffic management scene on each type of roadway and situation.

This book stresses safety practices during all phases of traffic management. The information contained in this book will help you practice good coordination, cooperation and communication -- an essential element when working with all responding agencies

Supplement to This Book

An *Instructor Resources CD* to accompany *Highway Scene Safety* is available, and includes many features to help instructors prepare and deliver classroom instruction on the content of the book:

- *Lesson Plans* detail the important points in each chapter, and correlate to the accompanying PowerPoint presentations included on the CD.
- *Answers to the Review Questions* in the book are included to help instructors validate student learning.
- *PowerPoint Presentations* outline the important points in each chapter and include photos and graphics to enhance classroom presentation and comprehension of the content.
- *Quizzes* offer instructors the opportunity to evaluate student knowledge of the content introduced in each chapter.
- *Incident Command Forms* found in the back of the book are also available electronically on the CD.

Instructor Resources CD Order#:1-4354-6977-1

About the Author

Les J. Greenwood has 32 years of emergency services work in and out of New York State. Mr. Greenwood was the Fire Chief of the Chester Fire Department, a State Fire Instructor, a Fire Investigator, and a Police Officer overseeing the training division, and is currently employed with the New York State Office of Fire Prevention and Control. In addition, he is a member of the NYS Incident Management

Team as Planning Section Chief, and was assigned to disasters such as the World Trade Center, Hurricanes Dennis and Gustav, and several wildfires. He is a certified National Incident Management System instructor for the state and serves as a Deputy Coordinator for Special Operations for Orange County Division of Fire. He is also a member of the I-95 Coalition Tri-State Highway Operations Group for the Lower Hudson Valley in New York and on the NYS Steering committee for Highway Safety.

Acknowledgements

Highway Scene Safety was made possible by some dedicated individuals that have contributed to its making. Thanks is given to the following individuals and agencies:

Joseph Andre
Chief, Mechanicstown Fire Department

Nick Elia
Asst. Chief, Mechanicstown Fire Department

Chris Whitby
Office of Fire Prevention and Control
Town of Wallkill Police Department, Wallkill, NY

Also, to those that I have not personally mentioned above, know that I appreciate any and all assistance and guidance that was offered to me.

The publisher and I would like to extend our appreciation to the reviewers who provided guidance in the development of the manuscript:

Ronald Eck
West Virginia University
Morgantown, WV

Clayton Lutz
Alderson Broaddus College
Philippi, WV

Philip Oakes
Casper College
Casper, WY

Tom Ruane
Fire and Life Safety Consultant
Peoria, AZ

I would also like to thank the person who started this venture in the Highway Safety issue, John Horan, Orange County Fire Coordinator. Without his foresight and encouragement this would not have been possible.

In addition, I would like to thank my wife, Patti, and son, Jason, for their patience, encouragement, and support throughout the entire process of writing this book. Due to their constant reviews of this book, both have displayed their knowledge and expertise to critique highway incidents!

The Importance of Safety at Highway Incidents

Learning Objectives

Upon completion of this chapter, you should be able to:

- Understand the necessity of training emergency responders regarding the safe setup of highway incidents.

- Explain the different laws, rules, and regulations regarding highway safety.

- Understand how vehicles react on different types of roadways.

Case Study

A volunteer Assistant Fire Chief and a Sheriff Deputy were fatally injured after being struck by a tractor-trailer while assisting at a motor vehicle incident caused by near-zero visibility conditions. Unforeseen weather conditions caused smoke from a contained fire on a nearby military range in conjunction with fog to move across a four-lane highway. A truck driver, attempting to slow down his tractor-trailer after encountering the smoke and fog, swerved suddenly to miss a vehicle parked in the highway. The tractor-trailer first struck the Sheriff Deputy's patrol car before striking and killing the Assistant Fire Chief.

The investigation determined that the key contributing factors causing the fatalities and injuries were the inability to establish traffic control on both sides of a divided highway, the ineffective coordination of the multiple agencies involved in the emergency response, and the unsafe vehicle operation of motorists during inclement weather and environmental conditions.

Investigators from the **National Institute for Occupational Safety and Health (NIOSH)**, the governmental agency responsible for conducting research and making recommendations for the prevention of work-related illnesses and injuries, made recommendations based on their investigation of the incident. NIOSH concluded that fire departments, EMS, and municipalities should take the following specific measures to minimize the risk of similar occurrences:

Fire Departments
- Ensure that first responders to a highway incident control oncoming traffic before responding to the emergency.
- Establish pre-incident traffic-control plans and pre-incident agreements with law enforcement and other agencies such as highway departments.
- Ensure that high-visibility vests meet the minimum requirements of the **American National Standards Institute (ANSI)** and **International Safety Equipment Association (ISEA)** standards, ANSI/ISEA 107-2004 or ANSI/ISEA 207-2006. ANSI is a private,

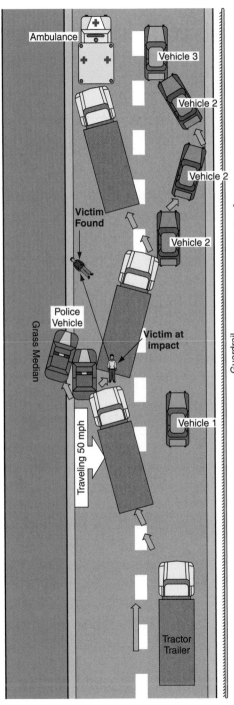

not-for-profit organization promoting a voluntary consensus of standards for products to ensure safety and health. ISEA is trade association whose member companies promote the protection of health and safety for workers.

EMS

- Ensure that emergency vehicles are parked in protected work areas when responding to emergency situations.

Municipalities

- Consider establishing a multi-agency communication system for response operations to coordinate and communicate incident activities.

More information on this incident can be found at: http://www.cdc.gov/niosh/fire/reports/face200817.html.

▨ INTRODUCTION

Safety for all responders—regardless of service field—is one of the most important issues concerning the emergency services. During any operation involving law enforcement, fire service, or emergency medical services (EMS), the primary concern of the incident commander is safety for all responders. The case study clearly illustrates how a response can quickly become dangerous under inclement weather and environmental conditions. Without safety in mind, somehow, somewhere, a responder will either be injured or become another fatality.

Looking at injuries that have occurred during highway incidents requires us to consider our operations and ask ourselves some serious questions:

- Is there a safer method for completing the objectives? (This question should be asked throughout every operation.)
- Do we need to change our operating standards to safely execute a specific traffic stop, collision scene, or another operation?

If the answer to either question is "Yes," then we need to revise our methods to ensure safety for all responders.

Cooperate, Communicate, and Coordinate

With the previous example of a struck-by incident in mind, we need to ask ourselves, "Did we do everything possible to follow operating procedures and control traffic prior to putting members in harm's way?" This question needs to be on the incident commander's mind when responding to and operating at a highway incident.

We are all responsible for safety. As you read further, you will see what can be done to keep everyone safe and how we can cooperate, communicate, and coordinate to send everyone home safely:

- **Cooperation** with other agencies that you work with on a regular and even an irregular basis is necessary. Preplanning with agencies prior to having an incident promotes cooperation by instilling confidence in the ability of one agency to help another under stressful situations.
- **Communications** is essential to every incident. For example, making sure that you and the police agency responsible for the incident have an effective and understandable flow of communication—whether by radio or phone. Without effective communications, your rescue effort could be hindered.
- **Coordination** is essential for all three "Cs" to work. Without good coordination, everything could slow down and even disrupt essential services from getting to the scene when needed.

Governing Laws, Rules, Regulations, and Standards

In the last six years, we have come to enforce certain laws, rules, and regulations to assist with keeping our highway incidents safe. Keep in mind that rules have been developed as a result of previous incidents that have occurred.

The Manual on Uniform Traffic Control Devices (MUTCD)

The *Manual on Uniform Traffic Control Devices*, or **MUTCD** (see Figure 1-1), defines the standards used by road managers nationwide to install and maintain traffic control devices on all public streets,

Figure 1-1 MUTCD.

highways, bikeways, and private roads open to public traffic. The MUTCD is published by the Federal Highway Administration (FHWA) under 23 Code of Federal Regulations (CFR), Part 655, Subpart F.[1] This federal manual was the first to mention emergency services operations on today's highways. In particular, section 6I references emergency services operations and how they should be protected.

Many sections of the MUTCD—used by the Department of Transportation (DOT)—also make sense in emergency operations. These areas include the use of **temporary traffic control zones**, chevron-shaped reflective tape placed on the rear of an apparatus, and buffer vehicles placed strategically at the scene. A temporary traffic control zone used to modify the flow of traffic near a highway incident should be implemented by using signs, signals, or markings that conform to the standards of the MUTCD.

Construction zones (see Figure 1-2) have been set up by construction workers for safety since the first roadway was built. Construction

© Delmar/Cengage Learning.

Figure 1-2 Photo of Construction Zone.

workers have experience in the art of managing traffic on a daily basis. Some of their tactics and equipment can also be used by emergency services. Using traffic vests and traffic warning signs and controlling traffic patterns can help improve safety for highway-incident responders. Additionally, working in a safe environment can generate greater productivity for responders.

The use of reflective vests while working around traffic is essential to the safety of responders on the highway. The use of pink-colored signs and the specific setup of cones help notify motorists of temporary incidents on the roadway so they can take appropriate action. The MUTCD also mandates the use of the Incident Command System (ICS), as outlined in the federal guidelines in the National Incident Management System (NIMS). These important safety measures are outlined in this manual to keep highway-incident responders safe and bring agencies together for a safer scene.

National Fire Protection Association (NFPA)

All these MUTCD operations have been recently incorporated into the **National Fire Protection Association (NFPA)** standards *NFPA 1901 Standard for Automotive Fire Apparatus and NFPA 1500 Standard on Fire Department Occupational Safety and Health Program.* The NFPA is a nonprofit organization that develops, publishes, and disseminates codes and standards intended to minimize the possibility and effects of fire and other risks.[2] *NFPA 1901* and *NFPA 1500* have been the standards of care used for emergency services for some time now. *NFPA 1500* includes the setup of temporary traffic control zones and the use of traffic vests. Since January 2009, *NFPA 1901* has mandated the use of reflective chevrons on the rear of an apparatus, the placement of traffic cones on an apparatus, and the placement of traffic vests on every riding position in all new apparatus.

Move-Over Law

One safety standard that has been enacted into law is the **move-over law**. This law requires motorists traveling on multi-lane roadways to—when practical—merge away from a vehicle working on the side of the highway to provide an empty travel lane of safety for workers. If not practical (either due to traffic volume or road design), the motorist must slow down significantly below the posted speed limit while passing roadside workers.

The move-over law has been adopted by nearly every state to keep emergency responders safer on roadways. This law and the previously discussed standards are keeping emergency responders safer every day.

Directing Traffic, Closing Roads, and Rerouting Traffic

Always be aware of the laws in your state when it comes to directing and rerouting traffic or closing roads to traffic. Follow the policies and procedures of your jurisdiction. Rerouting traffic could create additional complications to your overall traffic plan and may add to traffic congestion within your jurisdiction. For example, consider low bridges and weight limits on roads inside a village coupled with the turning radius of large vehicles. If you send a large commercial vehicle through your village and you have a bridge with low clearance, would the vehicle you sent make it under the bridge or get stuck and create further traffic restrictions? When you consider your traffic plan in your preplanning meetings, bring in the local highway department or county GIS (geographic information system) that could map out your intended routes with restrictions. This is discussed further in Chapter 2.

Safety Concerns at a Roadway Incident

Safety protocol should be developed and instilled into emergency responders prior to operating at a roadway incident. These issues are:

- Dismounting from an apparatus (see Figure 1-3)
- Walking around the protected area vs. the lanes of travel
- Always being aware of your surroundings
- Controlling others, such as curious onlookers
- Controlling the movement of the media

Figure 1-3 Prior to Dismounting Apparatus, Check Traffic.

Vehicle and Driver Reactions on Roadways

Traffic management depends, in part, on how vehicles may react given certain situations, such as roadway surfaces. Roads can be constructed from many different materials and contain characteristics that may affect a driver's ability to maintain control of the vehicle.

Depending on the time of year or even the time of day, road surfaces can vary. During cold weather conditions, the road surface can change quickly. Ice, sleet, snow, and/or rain can change how a vehicle will react.

All factors should be taken into consideration during the preplanning process with each agency. Vehicle reactions will determine the safety of all aspects of a highway-incident operation. Remember not only how vehicles might react under ideal conditions but also how they might react under adverse conditions.

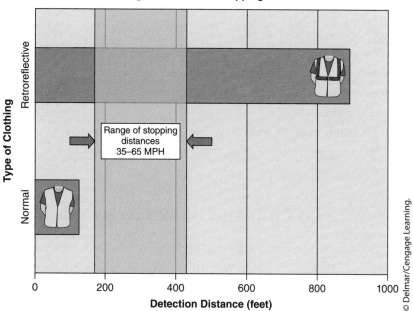

Figure 1-4 Stopping Distances.

Reduced Visibility

During or after a storm, fog in the morning, snow and rain, dust, sun, and smoke create a reduced visibility scene. A driver's ability to see what is ahead will be limited. Additionally, the time of day that a highway-incident occurs may affect visibility. For example, incidents occurring at night may have reduced visibility.

Reduced Traction and Steering Control

Drivers depend on good road conditions to stop. But snow, ice, and rain can reduce traction and steering control. The stopping distance necessary for a vehicle may be significantly lengthened (see Figure 1-4).

Roadway Materials

Every agency should be aware of the different types of materials used to construct roadways in their response districts. The roadway materials can affect the response time of drivers and vehicles. The following are examples of types of roadway materials:

- Tar and chip roads
- Blacktop roads
- Dirt roads
- Concrete roads

Tar and Chip and Blacktop Roads

New roadway surfaces, such as tar and chip or blacktop, create special concerns. These new roadway surfaces have stopping distances that are much greater than older roadway surfaces. Additionally, wet tar on a road during a hot summer day can be just as slippery as ice during the winter.

Dirt Roads

On a dry day, dirt roads or unfinished roads also create special concerns. Dust and dirt can be kicked up, obscuring visibility, which affects emergency services safety. Therefore, steps such as reducing the speed of the traffic around incidents should be part of your sight safety plan.

Types of Roadways

The following terms are used in identifying different types of roadways:

- Interstates
- Collector roads
- Local rural roads and streets
- Ramps
- Climbing roads

Figure 1-5 Interstate.

Interstates

Interstates (see Figure 1-5) are limited-access highways and are generally inter-regional, high-speed, high-volume, and divided roadways with complete controlled access.

Collector Roads

Collector roads (see Figure 1-6) have a dual function. They collect and distribute traffic while providing access to abutting properties.

Local Rural Roads and Streets

Local rural roads and streets are primarily town, county, village, and city streets (see Figure 1-7). They constitute a high percentage of the highway mileage and provide access to abutting properties.

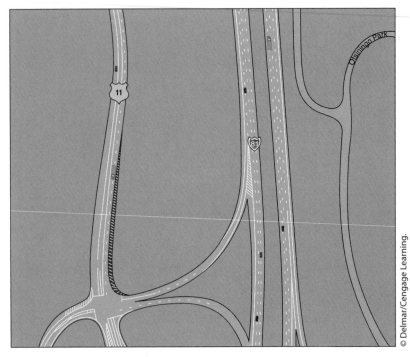

Figure 1-6 Collector Road.

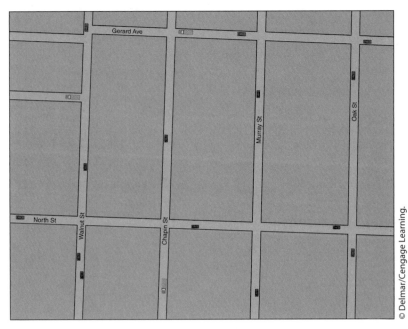

Figure 1-7 City Street.

Ramps

Ramps (see Figure 1-8) are turning roadways that connect two or more sections of an interstate. They can be multi- or single-lane roads, and they can be high-speed access. This constitutes careful attention when operating on or within them. These types of areas need to have special attention paid to them.

Climbing Lanes

Climbing lanes (see Figure 1-9) are auxiliary lanes provided for slow-moving vehicles ascending steep grades. They may be used along all types of roadways.

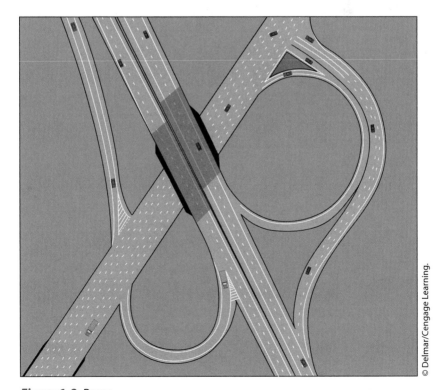

© Delmar/Cengage Learning.

Figure 1-8 Ramp.

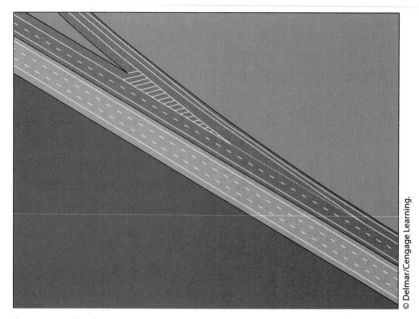

Figure 1-9 Climbing Lane.

SUMMARY

Emergency responders must understand the necessity of safety during highway incidents. The laws, rules, and regulations established by governmental and private nonprofit organizations help keep first responders at highway incidents safe. Additionally, understanding the uses of certain types of roadways, how they're built, and how vehicles will react to given roadway conditions helps keep everyone safe.

Review Questions

1. Name at least two types of materials used in constructing today's roads.
2. How do adverse weather conditions affect stopping a vehicle?
3. What type of roadway connects two limited access highways?
4. What's the primary consideration when it comes to setting up highway incidents?
5. What federal manual regulates temporary highway incidents?

Key Terms

American National Standards Institute
Climbing lanes
Collector roads
International Safety Equipment Association
Interstates
Manual on Uniform Traffic Control Devices
Move-over law
National Fire Protection Association
National Institute for Occupational Safety and Health
Ramps
Temporary traffic control zones

Endnotes

1. *Manual on Uniform Traffic Control Devices.* Retrieved from http://mutcd.fhwa.dot.gov on July 18, 2010.
2. National Fire Protection Agency. Retrieved from http://www.nfpa.org on July 18, 2010.

2

Traffic Management

Learning Objectives

Upon completion of this chapter, you should be able to:

- Explain the fundamentals of traffic control when working on the highways roadways? in your response area.

- Understand roadway characteristics and then apply them to setting up a safe work area.

- Explain the proper signage as it applies to temporary traffic control.

- Explain the importance of proper protective equipment.

Case Study

Two career firefighters were struck by a motor vehicle on a wet and busy interstate. One was killed and another was seriously injured. A ladder company and a squad car responded to a single motor vehicle crash that occurred on an interstate highway. After arriving on scene, the driver of the ladder truck parked near the median wall, approximately 150 yards directly behind the squad car, with the emergency lights turned on and operating, providing protection from oncoming traffic for both the occupant of the initial motor vehicle crash and the squad car personnel. Approximately 2–3 minutes after arriving on scene, the ladder truck was hit from behind by a motor vehicle. While firefighters attended to the injuries of the secondary accident, the Captain from the ladder company attempted to flag traffic away from the secondary accident.

At this time, the firefighter directing traffic near the original incident noticed a car cresting the top of the overpass on the inside lane rapidly approaching the secondary incident. He yelled two warnings over the radio. The car lost control, spun backward, impacted the median wall, slid between the wall and the ladder truck missing the Captain and one of the firefighters, before striking two other firefighters and the injured driver from the secondary incident. The two firefighters and the injured driver were thrown approximately 47 feet from the point of impact, killing one firefighter and seriously injuring both the other firefighter and the driver from the secondary accident.

The investigation of this incident produced the following safety precautions. All points are relative to this chapter and can help prevent such incidents from happening again.

- Establish and implement standard operating procedures (SOPs) regarding emergency operations for highway incidents.
- Ensure that the fire apparatus is positioned to take advantage of topography and weather conditions (uphill and upwind) and to protect firefighters from traffic.
- Ensure that firefighters responding to a scene involving a highway incident or fire must first control the oncoming vehicles before

Diagram 1

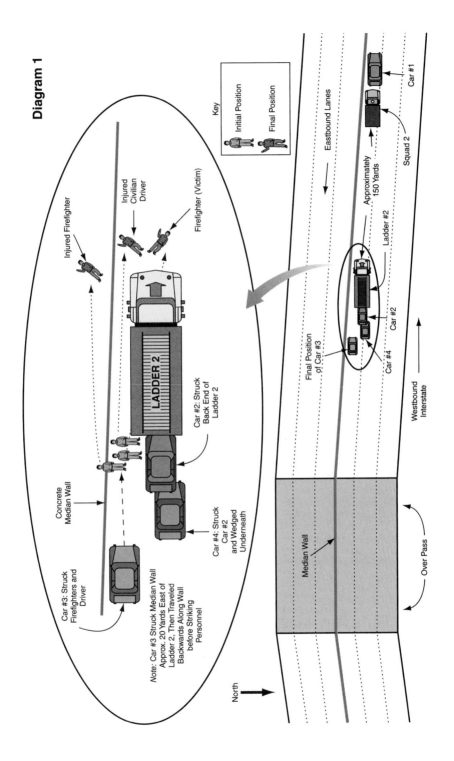

Key
- Initial Position
- Final Position

Eastbound Lanes

Car #1

Squad 2

Approximately 150 Yards

Ladder #2

Car #2

Final Position of Car #3

Car #4

Westbound Interstate

Median Wall

Over Pass

North

Injured Firefighter

Injured Civilian Driver

Firefighter (Victim)

Concrete Median Wall

Car #2: Struck Back End of Ladder 2

LADDER 2

Car #4: Struck Car #2 and Wedged Underneath

Car #3: Struck Firefighters and Driver

Note: Car #3 Struck Median Wall Approx. 20 Yards East of Ladder 2, Then Traveled Backwards Along Wall before Striking Personnel

21

safely turning their attention to the emergency in the event police have not arrived.

- Ensure that personnel position themselves and any victim(s) in a secure area, when it is impossible, to protect the incident scene from immediate danger.
- Use a "changeable message sign" to inform motorists of hazardous road conditions or vehicular accidents.[1]

Full report can be found on online at http://www.cdc.gov/niosh/fire/reports/face9927.html.

INTRODUCTION

Emergency services professions are dangerous. We live and breathe safety, but when we work in and around traffic, we sometimes forget how hazardous it is.

Some common misconceptions are:

- Drivers see our lights, and that's good enough.
- Wearing a vest isn't necessary because the fire apparatus provides enough warning.
- Traffic isn't a concern because police will control the traffic.
- The shoulder is a safe place to work because vehicles don't drive on the shoulder.

There are many more misconceptions, but hopefully, this chapter will give emergency service personnel an understanding of the need for protective areas and equipment to guide the motorists around the incident work area.

Fundamentals of Traffic Control

The *Manual on Uniform Traffic Control Devices* (MUTCD) includes a standard that we should all understand and use. The National Fire Protection Association (NFPA) has used these fundamentals in *NFPA 1001* and *NFPA 1500*. As the industry standard-setter, NFPA expects all fire services to become familiar with safety on the fire ground and to use the MUTCD to make highway work areas safer.

Traffic Incident Management Area

A **traffic incident** is an emergency road user incident, natural disaster, or another unplanned event that affects or impedes the normal flow of traffic.

A **traffic incident management area** is an area of a highway where temporary traffic controls are ". . . installed, as authorized by a public authority or the official having jurisdiction, in response to a road user incident, natural disaster, hazardous material spill, or other unplanned incident. . . ." Check with the local jurisdiction's policies and procedures to find out which agency is responsible for closing a roadway.

One of the primary goals when setting up a traffic incident management area is to control the traffic upon arrival. The first 10 minutes can be very hectic and stressful moments for everyone. The best way to control the situation is to stop traffic from entering the area until responders can make it safe. One way to easily and safely accomplish this task is to have an apparatus slow down traffic. This method avoids having a human responder from getting out of a vehicle to physically stop traffic. For example, while responding to an accident on a two-lane expressway, you can use two pieces of apparatus so one is in lane one and the other is in lane two. This way, traffic would slow down behind the vehicles.

SUCCESS STORY

One fire department recently purchased a new traffic control unit, which is specifically designed for cone setup and take down, arrow boards, lighting, and communications. Additionally, the department has a citizens band (CB) radio. While CBs seem outdated, the department has used this system with great success. With the cooperation of the commercial trucking industry, the fire department contacts the drivers of large vehicles by radio and solicits their help in slowing down traffic and sometimes stopping it when necessary. This approach has proven to be quite successful.

Temporary Traffic Control (TTC)

Temporary Traffic Control (TTC) means that the needs and control of all road users through the TTC zone are an essential part of highway construction, utility work, maintenance operations, and the management of traffic incidents. The reference to traffic incidents is the part of this standard that affects emergency services. See Section 6I of the MUTCD, which outlines the emergency services responsibilities with regard to TTC. Within this standard, the fire service must plan for incidents prior to response rather than during a response. This planning includes detours, scene assignments, and setup of TTC zones as well as the consideration of weather conditions, roadways, and length of time on scene.

On-Scene Size-Up

When arriving on the scene of an accident, car fire, or another roadway emergency, the first consideration is the safety of the responders. Without this, the entire operation is in danger. According to the MUTCD standard, when doing an on-scene size-up, responders are responsible for evaluating the magnitude of the incident, the expected time duration, and the expected **queue length**, or backed-up traffic. After completing this evaluation, responders should set up the TTC.

According to the MUTCD, during size-up, responders should use one of three categories to identify the incident:

- Minor
- Intermediate
- Major

Minor Incident

A **Minor Incident** is an incident that will take less than 30 minutes to remove the road user from the lanes of travel. An example would be responding to an emergency that requires EMS but where the vehicle is off the roadway.

Intermediate Incident

An **Intermediate Incident** is an incident that will likely take between 30 minutes and 2 hours. Intermediate incidents are the most typical incidents to which the fire service responds. These incidents are usually car fires or one- or two-car motor vehicle accidents.

Major Incident

A **Major Incident** is an incident that is 2 hours or more in length. This incident requires substantial time and resources to bring under control. Examples would be a multiple vehicle incident, a hazardous materials incident, a bridge collapse, or any other incident that would require extended operations.

The use of these three categories as a guideline can help you quickly size-up the magnitude of an incident. This allows for better planning, resource management, and safe working practices.

Standard Operating Procedures

NFPA has now outlined in the *NFPA 1500 Standard on Fire Department Occupational Safety and Health Program* what practices responders should follow when operating on roadways. Section 8, Traffic Incidents, states that all departments shall establish, implement, and enforce standard operating procedures (SOPs) when operating on roadways. Examples include setting up and using traffic cones, retro-reflective signs, illuminated warning devices, and other warning devices appropriate to warn oncoming traffic. Essentially, NFPA incorporates the MUTCD practices into *NFPA 1500*.

Having SOPs in place ensures good practices and good training for emergency personnel. You must enforce SOPs for each and every incident. Taking the time to set up a TTC greatly increases the safety of each and every responder.

Roadway Characteristics

The types and characteristics of roadways come into play when responding to an incident. Knowing a roadway's hills, curves, and type as well as construction sites on the roadway helps in both preplanning and on-scene planning of an incident. For example, an accident occurring just beyond a curve will have different demands than one occurring on a straightaway.

Responders need to know the following prior to responding to an incident:

- Location
- Type of roadway
- Speed of roadway
- Time of day

Location

Considering the location of the accident is of utmost importance. Whether the scene is on a straightaway or just after a curve helps to determine the location for the placement of the first warning sign. If an accident is after a curve, then responders place the sign just before the curve at a proper distance to ensure the safety and warning of on-coming traffic. The **line of sight**—shown in Figures 2-1 and 2-2—from which traffic comes is important when considering where to place cones and set up signs. Line of sight refers to the distance that emergency response personnel can see without any visual impairment or interference.

Responders may also place a flagger, police officer, or firefighter just before the traffic management area to support the management area and to serve as a lookout for traffic that disregards the traffic pattern. Such individuals would be of even more importance on

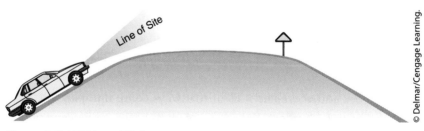

Figure 2-1 Hill Line of Sight.

© Delmar/Cengage Learning.

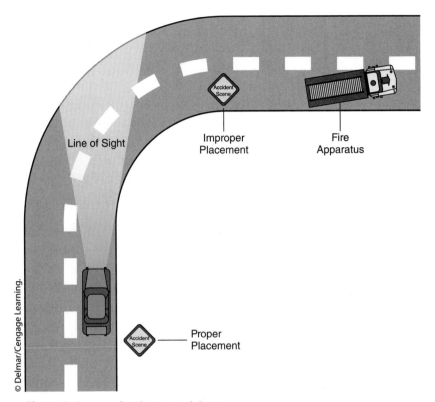

© Delmar/Cengage Learning.

Figure 2-2 Line of Sight Around Curve.

an expressway due to the high road speeds and characteristics of the road.

Type of Roadway

The type of roadway is important because surfaces and construction material have an effect on the vehicles that responders are trying to control. Some of the different roadway surfaces are asphalt, tar and chip, concrete, and dirt. As discussed in Chapter 1, each of these types of roadways has special concerns regarding stopping distances, kicking up dust and dirt, etc.

The stopping distances on these different roadways will affect how and where you will set up your traffic management area. For example, a newly asphalted road surface could be very slippery when wet; this needs to be taken into consideration when setting up your first warning sign.

> ### LESSON LEARNED
>
> A fire department shut down a roadway for an accident. The roadway was wet from a recent storm, and traffic slowed down, causing a backup that was getting longer by the second. A traffic warning sign was placed approximately 1,500 feet from the accident scene just prior to a hill on an expressway. Traffic naturally started to back up to the point where traffic was stopped just beyond the crest of the hill. Vehicles started getting out of control as they tried to stop, and larger vehicles nearly jackknifed to avoid a secondary accident. In order to prevent a second accident, traffic was allowed to move past the accident scene. The first sign was moved further up the roadway, which made the warning area longer.

Roadway Speed

Roadway speed is an important factor when planning and setting up the TTC. Expressways—shown in Figure 2-3—are much different than local roadways. The speed encountered on expressways needs to be controlled immediately in order to create a safe work area.

Time of Day

The time of day definitely comes into play when setting up the TTC. Imagine the differences between an accident that occurs at 4:00 p.m. and one that occurs at 2:00 a.m. Traffic volume—whether considering setting up detours or the response of an apparatus—is among the issues to review when planning a traffic control response to an incident. A good rule of thumb is that for every minute that traffic is held up, the result may be 1 mile of backup.

Figure 2-3 Interstate Traffic.

▨ Setting the Scene: TTC Zones

The components of a TTC zone—shown in Figure 2-4—are these essential areas:

- Advanced warning area
- Transition area
- Activity/work area
- Termination area

Advanced Warning Area

The **advanced warning area** is where motorists are first informed about the upcoming traffic pattern in your emergency work zone. Typical warning distances are shown in Table 2-1. For example, working on an expressway, responders would place the first warning sign at least 2,640 feet prior to the work area. As the speed of the roadway decreases, so does the distance of the first warning sign.

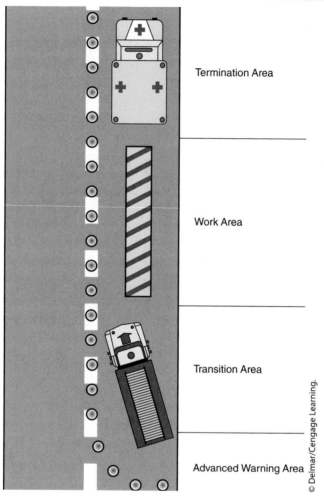

Figure 2-4 Example of a TTC Zone.

Transition Area

The **transition area** is where traffic is diverted from the normal path of travel. This area tells the motorist that an action is required because of a lane shutdown, a lane diminished, or another impending action. Traffic cones are typically the most used tool for this job. A buffer vehicle should be placed in the transition area, as shown previously in Figure 2-4.

Table 2-1   Advanced Warning Sign Distances. © Delmar/Cengage Learning.

Roadway Type	Distance between Signs		
	A	**B**	**C**
Urban (low speed)	100 feet	100 feet	100 feet
Urban (high speed)	350 feet	350 feet	350 feet
Rural	500 feet	500 feet	500 feet
Expressway/Freeway	1,000 feet	1,500 feet	2,640 feet

A = the distance from the transition zone to the first sign
B = the distance between the first and second signs
C = the distance between the second and third signs; the third sign is the first in a
    three-sign series encountered by a driver approaching a TTC zone.

Activity Area

The **activity area** is where emergency service work actually takes place. This area must be designated by the incident commander and protected for safe operations. The area could be short or it could stretch out for a long distance. A buffer space is included in the activity area. This buffer space should be used to place a blocker vehicle for the personnel on the scene. Several different types of vehicles can be used for this detail. The preference is to use a large vehicle, such as a truck, tanker, engine, etc. Using a smaller vehicle, such as a command or police vehicle, is not suitable for stopping or slowing down oncoming large vehicles and some types of passenger vehicles.

Termination Area

The **termination area** is the end of the area in which the emergency services operate. This area is where traffic returns to its normal path of travel and speed. Where the termination area ends determines if traffic control is needed. For example, if this area ends near of an off-ramp, responders may want to consider posting a flagger in the area.

© Delmar/Cengage Learning.

Figure 2-5 Landing of Helicopter and Placement of LZ is Important Prior to Landing.

Medivac Helicopters

In addition to the other areas to be determined at a roadway incident is what to do with Medivac helicopters coming to the scene. The normal operating procedure has been to close the roadway and land the helicopter. However, the roadway may not be the safest place to land. Responders should consider landing helicopters in predesignated landing areas, such as alongside an expressway.

Determining TTC Zones

Setting up a TTC zone is critical. Responders must take into consideration where on the roadway to set up this area. One place would be on a straightaway. The preferred method of placing the blocker vehicle to create a greater blocking area is to angle the vehicle.

Placing the staging area at the end of the incident gives easy access for EMS to package patients and leave with virtually no disruption to the traffic flow.

Scene Documentation

Refer to Appendix C for a checklist of items needed to make a scene safe and to provide accountability. For example, proper size-up will provide documentation for cone placement carried out during the incident. This document can be referenced for weeks, months, or years later if necessary.

Lane Numbering

Fire departments should discuss lane numbering with EMS and police for consistency in numbering lanes on expressways. Lanes should be numbered from left to right, starting with the number one, as shown in Figure 2-6. This procedure ensures that all responding agencies speak the same language when referring to lane numbers.

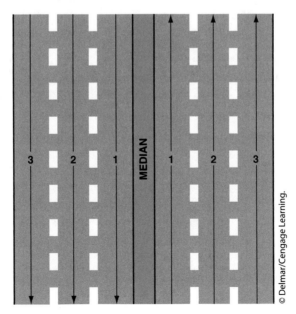

Figure 2-6 Graphic of Lane Numbers.

▓ Proper Signage and Equipment

According to the MUTCD, emergency services are to use the color pink and specific-sized signs to control traffic on roadways. The color pink has been chosen to represent temporary traffic control warning signs as specified in MUTCD. Each emergency services agency should evaluate the wording used on signs to control traffic in their specific jurisdiction. Some examples include:

- RAMP CLOSED
- LANE CLOSED (with directional overlays)
- FLAGGER AHEAD
- EMERGENCY SCENE AHEAD (recommended by *NFPA 1500*)

These signs give traffic more specific directions and can mean safer work zones. See Figure 2-7.

Figure 2-7 Emergency Scene Ahead.

© Delmar/Cengage Learning.

Sign Requirements

According to the MUTCD, warning signs are required to be a minimum size depending on the speed of the roadway. Two sign sizes are available for use: 48 inches by 48 inches for expressways and 30 inches by 30 inches for lower speed roadways. *NFPA 1500* further defines this requirement as a 48-inch by 48-inch pink sign that must be approved by the Federal Highway Administration (FHA). It is also important to be able to store these signs so they do not take up a lot of room on the apparatus (see Figure 2-8).

Placement of Signs

The placement of these signs is determined by what type of circumstances you encounter. As previously covered, the type of roadway, the characteristics of the roadway, the time, the date, the weather, the

Figure 2-8 Storage of Signage.

traffic flow, and the location of the incident affect the placement of signs. Responders must consider all these conditions in order to make a sound decision when creating a work area and making it safe.

Portable Signs

Portable variable message signs—shown in Figures 2-9A and 2-9B—display changeable electronic messages and are available to incorporate on your vehicle and apparatus. Some departments place message boards on the rear of the apparatus to warn motorists. Others place portable message boards on the side of roadways to create a message to fit the specific needs of each individual incident. These signs are similar to the Department of Transportation (DOT) signs but are compact enough to fit inside a command vehicle or an apparatus compartment.

Other Resources

The DOT may have other types of signage available. One possibility is the numerous variable message boards already placed throughout the jurisdiction. An example would be portable message boards placed in high-activity areas. These signs can be accessed via cell phone by a DOT supervisor at any time day or night. These signs could be considered for use as an advanced warning for an accident scene. Each state is creating **traffic management centers (TMCs)**, the centralized location where traffic-related data are collected and from which all traffic-related information can be disseminated. These TMCs have dispatch capabilities to assist with creating messages on permanent or portable message boards. Planning in advance to use such tools can help create a safer work area for all responders.

Traffic Cones

Using cones as channeling devices is listed in the MUTCD. Cones with a minimum size of 28 inches and two reflective strips—one 4 inches in width and one 6 inches in width—are standard equipment and should be used when channeling traffic into desired lanes. These

(A)

(B)

Figure 2-9 A and B Variable Message Board.

© Delmar/Cengage Learning.

Figure 2-10 Arrangement of Cones Is Essential When Developing a Traffic Guidance Plan.

cones are to be used in transition areas as well as the activity area (see Figure 2-10).

Flagger Signs

The MUTCD has a standard for flaggers to use during any kind of operation. Because there are times that fire departments must control traffic with either fire police or traffic control units, responders need to have the proper STOP and SLOW paddles. Chapter 3 discusses flagger operations and the equipment needed to perform these functions properly on the roadway.

▓ Personal Protective Equipment (PPE)

Regarding traffic control standards, the MUTCD requires emergency services personnel to wear high-visibility safety apparel, as outlined in ANSI 107-2004. The use of standard NFPA firefighter gear, police officer uniforms, or EMS uniforms doesn't meet this standard when working on roadways. Firefighter protective equipment doesn't have the required daytime and nighttime reflective material to meet this standard.

Types of Vests

Currently, ANSI 107-2004 has three different types of vests:

- **Type I:** This type is for parking attendants, shopping cart workers, warehouse workers, and delivery drivers. Consideration is for vehicle and movement speeds of less than 25 mph (see Figure 2-11A).
- **Type II:** This type is for roadway construction workers, utility workers, survey crews, railway workers, forestry, parking, package handling, emergency workers, law enforcement, and accident scene investigators. Consideration is given to vehicles and movements of speeds greater than 25 mph and that have close proximity to vehicles; inclement weather is also considered (see Figure 2-11B).
- **Type III:** This type is for workers exposed to significantly higher vehicle speeds and reduced sight distances. This type should be used when a situation clearly places a worker in danger. The wearer must be comfortable in a full range of body motions at 1,280 feet and must be identifiable as a person. This would include roadway construction, utility, survey workers, flagging crews, law enforcement personnel, and emergency services workers (see Figure 2-11C).

Hazard Assessment for Vest Choices

The ANSI standard for vests doesn't dictate the type of vest that a department should use. Agencies must determine which vests are the safest by completing a hazard assessment of their jurisdiction. If an agency doesn't respond to expressway incidents, a Class II vest may be sufficient for its needs. However, if the agency responds to highway incidents, a Class III vest may better fit its needs.

Another consideration for PPE is the color of the equipment you will provide to your personnel. For example, if you have an apparatus that is safety yellow, you wouldn't choose high-visibility yellow as the color of

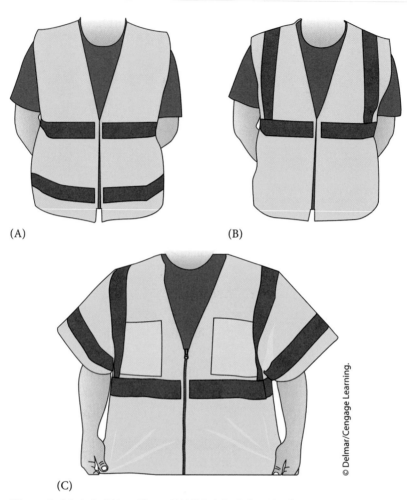

(A)

(B)

(C)

© Delmar/Cengage Learning.

Figure 2-11 A, B, C Vest Photo (ANSI Public Safety Vest).

choice. Orange or red would contrast better with this color of apparatus, as shown in Figure 2-12. Choosing the right background color and reflective color is as important as what type of vest you will need.

Additionally, ANSI has instituted the use of a new public safety vest for emergency services. This vest is not only designed to make the user visible but is designed with a tearaway feature, providing the wearer with a safe advantage. If it gets caught on close-moving vehicles, the vest is designed to tear off from the wearer so he or she isn't dragged down the road. See Figure 2-13.

Figure 2-12 Yellow Vest Against Yellow Vehicle vs. Orange Vest Against Yellow Vehicle. Notice Members at Right Are Visible from a Distance but the One Member on Left Is Barely Noticeable.

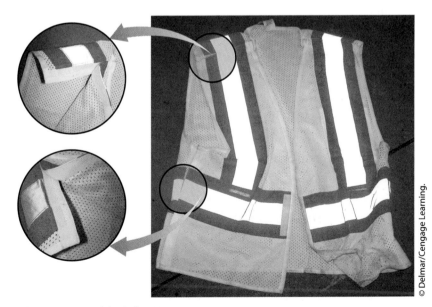

Figure 2-13 Public Safety Vest.

Some questions agencies should consider when choosing vests are:
- What color is our apparatus?
- What vest type is best for the type of incident to which we respond?
- Is the vest functional over our current PPE?
- Will the vest meet daytime as well as nighttime visibility requirements?

Breakdown of an Incident

If the incident scene were set up as previously suggested, responders would break down the incident by beginning with the last setup point and then work backward. They would go from the termination area back to the beginning of the traffic management zone. Depending on what has been set up, this process may need to be a coordinated effort among all agencies involved.

Generally speaking, it's much easier to stop all traffic before breaking down cones and signs. Responders can hold traffic until all apparatus and agencies are clear and then let the traffic continue with the original pattern. This method is the safest and easiest way of accomplishing this final operation.

Never discount safety as being too long to accomplish—it's everyone's responsibility. Having good situational awareness of your surroundings will keep you and your fellow first responders safe. Remember the 16 life safety initiatives of "Everyone Goes Home":
1. Define and advocate the need for a cultural change within the fire service relating to safety; incorporating leadership, management, supervision, accountability and personal responsibility.
2. Enhance the personal and organizational accountability for health and safety throughout the fire service.

3. Focus greater attention on the integration of risk management with incident management at all levels, including strategic, tactical, and planning responsibilities.
4. All firefighters must be empowered to stop unsafe practices.
5. Develop and implement national standards for training, qualifications, and certification (including regular recertification) that are equally applicable to all firefighters based on the duties they are expected to perform.
6. Develop and implement national medical and physical fitness standards that are equally applicable to all firefighters, based on the duties they are expected to perform.
7. Create a national research agenda and data collection system that relates to the initiatives.
8. Utilize available technology wherever it can produce higher levels of health and safety.
9. Thoroughly investigate all firefighter fatalities, injuries, and near misses.
10. Grant programs should support the implementation of safe practices and/or mandate safe practices as an eligibility requirement.
11. National standards for emergency response policies and procedures should be developed and championed.
12. National protocols for response to violent incidents should be developed and championed.
13. Firefighters and their families must have access to counseling and psychological support.
14. Public education must receive more resources and be championed as a critical fire and life safety program.
15. Advocacy must be strengthened for the enforcement of codes and the installation of home fire sprinklers.
16. Safety must be a primary consideration in the design of apparatus and equipment.

▥ SUMMARY

Responders need to understand the fundamentals of setting up, conducting, and breaking down a roadway incident. Many factors must be considered based on the category of the incident and the agencies responding. Good traffic control includes on-scene size-up to help responders effectively set up TTC zones, understand roadway characteristics, and determine proper signage for the incident. Additionally, to ensure the safety of responders, proper protective gear should be worn for every roadway incident.

▥ Review Questions

1. When setting up the traffic management area, where in relation to the work area should the blocker vehicle be placed?
2. Name the four sections of the traffic management area.
3. Upon arrival, the incident commander must size up the scene and decide the size of the incident. What are the three categories?
4. What color does the MUTCD suggest that emergency responders use for signage?
5. The Federal Highway Administration (FHA) has created a standard that regulates temporary highway incidents. What is that document called?
6. The National Fire Protection Agency (NFPA) has created a standard that directs the fire service to use specific signs on the highway. What is that standard?

Key Terms

traffic incident
traffic incident management area
temporary traffic control
queue length
minor incident
intermediate incident

major incident
advanced warning area
transition area
activity area
termination area
portable variable message signs
traffic management centers

Endnote

1. "One Fire Fighter Died and a Second Fire Fighter Was Severely Injured After Being Struck by a Motor Vehicle on an Interstate Highway—Oklahoma." Published December 30, 1999. A summary report for this incident is available at http://www.cdc.gov/niosh/fire/reports/face9927.html. Retrieved July 15, 2010. Case Study (NIOSH CASE 99F-27).

CHAPTER 3

Safety Officer Considerations

Learning Objectives

Upon completion of this chapter, you should be able to:

- Explain the responsibilities of a safety officer during a highway incident.

- Understand why early warning will help keep the work zone safer.

- Understand the importance of keeping the traffic flow moving in order to prevent secondary accidents from occurring.

- Explain why a quick clearance policy will help save responders.

Case Study

A motor vehicle collision disrupted traffic for miles. First responders on the scene were able to keep everyone safe and keep the flow of traffic moving by providing oncoming traffic with an advanced warning signal. Without advanced warning traffic flowing toward a highway incident, traffic often comes to an abrupt stop or slows down suddenly from traveling at high rates of speed. Traffic that is forced to suddenly stop or rapidly slow down from a high rate of speed is more likely to become involved in a secondary incident.

In this case, a state trooper used his own vehicle as an advanced warning to traffic. Along with advance warning signs, the state trooper drove his vehicle with the moving traffic and used his warning lights to help caution and slow down the vehicles in motion. Using his vehicle as an advance warning resulted in reducing the speed of traffic faster and more effective than advance warning signs alone.

Along with slowing down traffic, this single action includes added benefits:

- The trooper can keep the incident commander informed of traffic conditions.
- Traffic can be monitored for any secondary incidents that might not otherwise be as quickly discovered.
- Better communications can be maintained between the traffic control personnel and this advanced monitoring.
- Motorists will be kept on a higher level of alert than with the use of signage alone.

■ INTRODUCTION

The **safety officer** is responsible for all responders at an incident. When operating on highways, the safety officer must be in tune with what can happen when in the absence of certain safety considerations. The safety officer must work to keep the work zone safe while monitoring traffic backup and to prevent additional incidents from occurring.

▒ Safety Officer Responsibilities

The safety officer should determine whether to station a lookout at the beginning of the traffic management area (see Figure 3-1) in order to anticipate potential issues with traffic patterns.

The safety officer must also focus on safety issues that are created because of the traffic management area, considering the dangers prior to entering the area, and possibly placing a police vehicle or some prior warning device in that location.

Surveying the Scene

When surveying a scene, the safety officer should communicate with all involved agencies. This is one of the reasons we promote communication, cooperation, and coordination (the three Cs of incident management) as discussed in Chapter 1.

In addition, the safety officer should do the following:
- Monitor the **hot zone**, or work zone, of the traffic management area (see Figure 3-2). This is usually associated with any safety actions within the work zone.
- Make sure traffic management signage is properly placed.
- Make sure emergency response vehicles are properly placed and situated as dictated by agency policy and consistent with sound safety practices.
- Make sure all response personnel are wearing proper protective apparel.
- Make sure sound, common communications are used among all agencies.

When feasible, two safety officers should be assigned to a traffic management incident: one to operations and the other to scene safety. The **operations officer** should monitor what's happening within the hot zone. This position ensures that responders are following good, sound safety practices within the actual operation. The scene safety officer looks at the entire incident outside the work zone. For example, the scene safety officer would look at where the vehicles are placed,

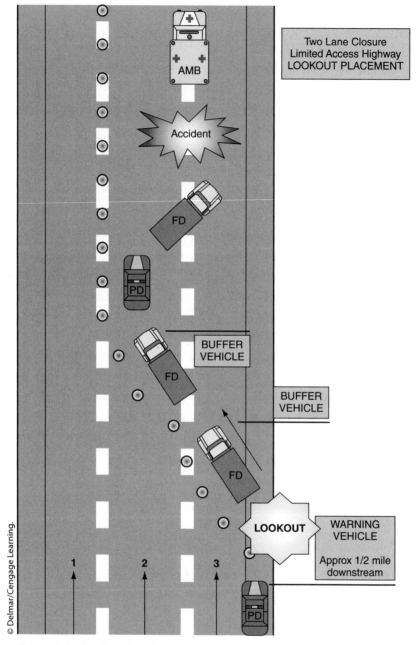

Figure 3-1 Graphic of Traffic Management Zone and Lookout Position.

© Delmar/Cengage Learning.

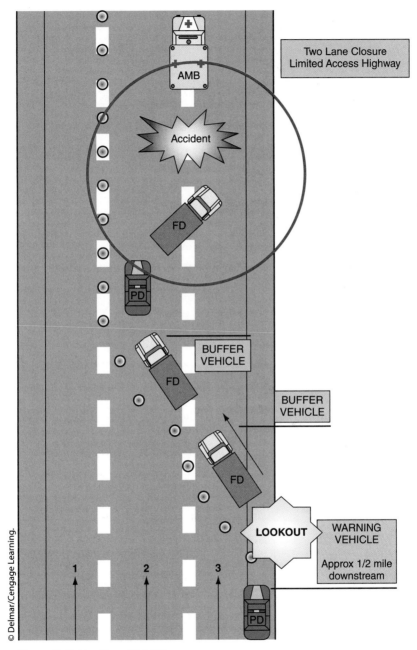

Figure 3-2 Hot Zone/Cold Zone.

Figure 3-3 Crash of Apparatus.

how well the traffic is flowing, and provide for an early warning of vehicle penetrations of the entire traffic management zone to avoid disasters such as the one shown in Figure 3-3.

The safety officer should be equipped with an early warning device, such as a loud horn (see Figure 3-4). One option is a canister of air designed to be a loud horn. This canister may be obtained at any sports store as a typical canister of air with a horn on top. When a safety officer—or lookout—sees a rogue vehicle, the horn would be sounded three times, signaling an immediate hazard to the work zone. The signal would be followed up by radio communications to personnel on the scene.

Another option is to have a radio signal similar to a mayday call in a building. Saying a term such as "runner, runner, runner" on the radio could signal a rogue vehicle that might endanger the responders.

Whatever signal is used, each agency needs to make this part of the policies and procedures for training all personnel. This training should not only include one agency but all who may respond with

Figure 3-4 Loud Air Horn.

your agency so everyone will recognize the safety hazard associated with this signal.

Responder Early Warning

An early warning for potential injuries to responders is key to making the scene safe for all to operate. If a department develops an entire incident according to safety standards recommended in the MUTCD but when no one is looking a vehicle runs through the cones and signage setup, *this creates a substantial safety issue!* Without anyone on the scene to alert other responders, the vehicle becomes an immediate danger to other drivers as well as those working on scene. Posting a safety person at the beginning of the queue to warn responders can help avert potential deadly injuries.

Having sufficient warnings in place needs to be part of the planning process and the development of standard operating guidelines of

Figure 3-5 The Coordinated Use of Communications by the Incident Commander is a Key Component of Any Incident.

all departments that will be operating on our highways. Law enforcement officers or firefighters can be stationed to observe what the traffic is doing during the incident.

One other consideration for the safety officer is to have proper communications with the incident commander (see Figure 3-5), which can be accomplished by planning a common frequency for communications of safety issues. This communication should be part of the preplanning meetings with agency heads.

▧ Traffic Rehabilitation Concerns

One of the least looked-at areas of a traffic incident is **rehabilitation**—getting the traffic flow moving again—during long-term operations. During extended operations, it's possible to be consumed by the cause of the incident at hand. Safety officers must rehabilitate a traffic stoppage during a long-term incident (2 or more hours). The traffic behind

© Delmar/Cengage Learning.

Figure 3-6 The Continued Back up of Traffic at an Incident can Create Secondary Incidents with Motorists Using the Emergency Cross-over.

an incident is backed up on the highway, and motorists have nowhere to go (see Figure 3-6). Without guidance, drivers will often cross over lanes and potentially cause a secondary incident in the oncoming lane(s).

When traffic backs up, several issues may need to be addressed:
- Elderly sitting in traffic
- Medical situations created by sitting in the backed-up traffic
- Feeding vehicle occupants
- Vehicle issues: flat tires, fuel issues, etc.
- Secondary crashes

Secondary Crashes

A **secondary crash** (see Figure 3-7) commonly occurs when the initial crash causes traffic within the lanes to back up and create congestion.

All incident management personnel should be very aware of the possibility of secondary crashes. This is often a primary cause of friction among the response personnel at the scene. Law enforcement

Figure 3-7 The Use of Advanced Warning can Reduce Secondary Incidents.

is always aware of the benefit of moving traffic to reduce the risk of secondary crashes. Secondary events can bring more apparatus and personnel to the scene, translating into more confusion, more time to clear the incident, and greater confusion to the traveling public.

A secondary crash is only one part of the highway safety equation, but we must realize that it's an important factor to consider in the overall size-up of any highway incident.

Incident within an Incident

One tactic is to assign law enforcement to patrol the area in order to keep the public secure and feeling calm while they're in a position in which they have no control.

Even with an extra person assigned to the traffic backup, the traffic backup can at times become an incident itself. This is commonly called an **incident within an incident** by the emergency management community.

LESSON LEARNED

A major traffic incident occurred during a significant snow-storm. This traffic incident caused a large backup of traffic on the interstate, leaving trucks, vehicles, and occupants stranded on an unplowed roadway. This situation caused great concern for the incident commander due to the limited access to backed-up traffic, stranded motorists in vehicles, possible exposure to extreme cold by vehicle occupants, and the lack of DOT access. Additional services had to be called and assigned to the second-ary incident and managed along with the initial traffic call. This situation placed more strain on the original incident, along with extended planning and logistical support. For example, extra ambulances, law enforcement officers, the possibility of needing to feed stranded motorists, and all other factors were involved. Safety concerns with the secondary incident soon became the primary focus of the incident commander, which necessitated the assigning of an additional safety officer just to handle the secondary event.

Quick Clearance Policy

The **quick clearance policy** is rapidly becoming the basis for reducing traffic incident injuries and fatalities among emergency responders. A quick clearance policy is one that promotes fast cleanup of the traffic incident to reduce the traffic backup as soon as possible.

The problem that quick clearance addresses is the last step: the **recovery process** (see Figure 3-8). Sometimes, this step can be longer than the original incident.

Bringing all agencies on board with quick clearance will reduce the time of exposure for all involved. Even though the primary stake-holder is law enforcement, informing all agencies regarding how quickly clearance can transpire helps keep all safe on the roadway.

Figure 3-8 Recovery Process.

Clearing an incident may start the moment an incident occurs, which creates traffic shock waves, with a slow-moving queue going upstream until all traffic is moving in a normal manner. Therefore, motorists experience congestion but never see any apparent reason for it. By the time the motorists get to the incident site, the incident is already cleared and all agencies have left the scene.

Reducing the time on scene, especially with rehabilitation, we can reduce the backup of traffic flow and emergency services' exposure to "struck-by" incidents.

SUMMARY

The safety officer must maintain a safe working area for responders while monitoring the possibility of secondary incidents relating to rogue vehicles and traffic backup. Safety officers need to keep traffic flowing and adopt quick clearance policies when possible to keep everyone safe.

Review Questions

1. Having two safety officers will greatly increase the safety of your operations. What are the two specific assignments for these safety officers?
2. What is a secondary accident?
3. What is an incident within an incident?
4. How does traffic backup for long periods of time create a hazard for emergency responders?
5. How can having a quick clearance policy help emergency responders?

Key Terms

safety officer
hot zone
operations officer
rehabilitation
secondary crash
incident within an incident
quick clearance policy
recovery process

CHAPTER 4

Flagging Operations

Learning Objectives

Upon completion of this chapter, you should be able to:

- Understand the reason for training in flagging operations.

- Conduct a proper flagging operation.

- Practice safe measures while conducting flagging operations.

- Understand the proper communications needed for flagging operations.

- Explain the procedures to prevent a flagger emergency.

Case Study

Firefighters assigned to flagging operations for a highway incident on a busy two-lane highway discovered that performing flagging operations safely was made difficult by drivers on the roadway. A driver who failed to see the flag narrowly missed hitting one of the flaggers, overcorrected, and drove off the road, causing a one-car secondary accident. The secondary incident could have involved a flagger injury or fatality if the flaggers had not been conducting safe flagging operations and acted quickly to get out of the way of a distracted driver.

Drivers can be very polite and cooperative, demanding and difficult, or distracted and dangerous. In order to keep traffic moving, flaggers must master the art of giving quick directions, maintaining patience, and being alert to the sometimes unpredictable actions of drivers.

Safe flagging operations are critical for the following:

- Ensuring the safety of personnel responding to a highway incident
- Reducing the workload on command at a highway incident
- Keeping traffic moving while responders conduct accident scene operations
- Alerting traffic that an incident has occurred
- Ensuring the safety of the general public driving on the roadways
- Helping prevent secondary incidents caused by inattentive drivers
  These are just a few of the reasons that we need to educate our emergency responders regarding safe flagging operations.

▓ **INTRODUCTION**

Because flagging operations is an important aspect of first responders, we must train our personnel to meet performance standards as part of the Occupational Safety and Health Association (OSHA) standard. However, the greater reason than meeting a regulation is to keep responders safe on the roadway. Training each and every first responder to regulate traffic gives the organization flexibility. Even when the assigned flagging personnel aren't available, any member can take the flagging position.

▓ Safety for Responders

Safety is the most important reason for training. Imagine a first responder taking a hose line during his first 24 hours and trying to put a fire out. Fire departments wouldn't even consider putting an untrained first responder in that position. Departments should train for flagging or traffic control for the same reason. When training first responders in the area of flagging, the following issues need to be addressed:

- Safety of the traffic management area
- Safety of the flagger
- Consistent signals from flaggers
- Placement of the flagging station
- Placement of traffic management signs

Remember that safety of the flagging operators and the mission is paramount.

Communications

All personnel in the traffic control division of an incident should work in a coordinated effort. Flaggers should able to see each other on the roadway and have the proper communications (see Figure 4-1). Radio communication between flaggers is mandatory and essential. Communications equipment shall include a radio with case. This will enable the command staff to communicate any orders they may have. Also, the flagger will be able to communicate with other posts and the command for any safety issues that may arise during this detail.

▓ Conducting Proper Flagging Operations

As with all emergency services, training consistency is of the utmost importance. Each person assigned to this detail must be trained in the same manner. The agency has to develop standards for signals, personal protective devices, and ensure the safety of all involved.

Figure 4-1 Being Visible on the Roadway Is Essential to Safety of the Member Working on the Highway.

Flagger Qualifications

Meeting the qualifications for a flagger is very important. Qualifications should include:

- Good physical fitness
- Good attitude and patience
- Mental alertness
- Courteousness
- Authoritativeness
- Professional appearance

Because a flagger is often the most visible responder, a good flagger must have all the above to make the traveling public feel that the emergency services have their best interest in mind (see Figure 4-2).

Professional Appearance

A good physical appearance is essential to any traffic detail. The flagger's appearance should instill confidence—rather than suspicion—in drivers. All emergency services workers must have a professional appearance each and every time they work in the public eye.

Figure 4-2 Well Equipped and Professional Looking Flagger.

Courteous and Authoritative Manner

In order for the public to cooperate and know that flaggers are professionals, flaggers must be courteous and authoritative at the same time. This is easy to say but sometimes hard to do. However, this approach to interacting with the public should be mastered. If flaggers act timid and unsure, the motorist will not respect the directions given and may choose to disregard the instructions. Being professional will enhance your authority and will encourage compliance, ensuring the safety of the general motoring public.

Flagger Operations Training

Each flagger must be trained in how to work in and around traffic. At any moment, a flagger may need to find an escape route to safety. Consider the following objectives when training personnel for flagging:
1. How to protect themselves (find an escape route)
2. How to flag correctly (all flaggers shall use the same signals and signage for each station)

3. How to protect the workers in the traffic management zone
4. How to maintain the physical ability to perform actions needed for safe, efficient operations during differing types of weather and conditions

In addition, the MUTCD states that flaggers shall be trained in the following areas:

- Ability to receive and communicate specific instructions clearly, firmly, and courteously
- Ability to move and maneuver quickly in order to avoid danger from errant vehicles
- Ability to control signaling devices (such as paddles and flags) in order to provide clear and positive guidance to drivers approaching a TTC zone in frequently changing situations
- Ability to understand and apply safe traffic control practices—sometimes in stressful or emergency situations
- Ability to recognize dangerous traffic situations and warn workers in sufficient time to avoid injury

Flagging Equipment

The equipment needed for the operation of a flagger station can be the most important aspect of the flaggers operation.

Signage

Consider what type of **signage** will be used. The following are some samples of signs:

- FLAGGER AHEAD (see Figure 4-3)
- LANE CLOSED
- RAMP CLOSED

Depending on your response area, signs may be needed to warn or direct motorists. Keeping the motoring public informed and aware of your presence is important to the safety of the flagger and the overall scene.

© Delmar/Cengage Learning.

Figure 4-3 Flagger Ahead Sign.

Other Flagger Equipment

Other essential devices that should be issued to and used by the flaggers are:

- Traffic safety vest
- STOP/SLOW paddle
- Lighting (used at night to light up the flagger area)
- Identification

Flagger Location

Flagger location—the placement of the flagger—and whether the flaggers will be able to see one another are important considerations. These decisions are part of a good communications plan. Developing the

Table 4-1 Table 6E-1 from MUTCD

Stopping Sight Distance as a Function of Speed			
Speed* (km/h)	Distance (meters)	Speed* (mph)	Distance (feet)
30	35	20	115
40	50	25	155
50	65	30	200
60	85	35	250
70	105	40	305
80	130	45	360
90	160	50	425
100	185	55	495
110	220	60	570
120	250	65	645
		70	730
		75	820

communications plan early will put any incident on the path to success. The terrain of the roadway, type of roadway, and curvature of the roadway should be considered to determine the location of the actual post. Responders must give motorists time to notice that there's an incident ahead and anticipate that they will encounter a flagger ahead. Therefore, placement of signage is an important part of this configuration.

Flaggers need to be far enough from the incident for approaching motorists to have sufficient time to stop. For example, on a roadway with a speed of 55 mph, responders will need the flagger station at a distance of 495 feet from the incident (see Table 4-1). This distance will provide a good traffic management zone.

Curves

Putting the flagger on the curve of a road will likely cause a secondary incident. A curve is an unsafe placement for a flagger station. If responders are operating on the curve of a roadway and need to deploy traffic control for the incident, they'll need to place a flagger on both sides of the curve. Placing flaggers in view of the traffic rather than the accident scene puts them in better control and a safer position and also makes room for traffic flow around the incident.

Hills

Similarly, when responders are on the down slope of a small hill—where traffic can't see them prior to cresting the hill—place the traffic control signs and flaggers prior to the crest of the hill. This way, the motorists are forewarned and the flaggers can maintain control of the traffic, which creates safer operations for all involved.

Flagging Equipment

Flaggers working on the roadway during emergency operations don't have the luxury of working under ideal situations. The emergency services are called into action no matter what the weather is. Because responders operate under less-than-ideal situations, they must protect themselves with proper equipment.

Vests

The flagger's **safety vest** (see Figure 4-4) is one of the primary safety items issued to a flagger. According to the MUTCD, flaggers must wear a vest to protect themselves from traffic. The reason for wearing a safety vest is to be seen. Different vests are required based on the incident conditions. As previously discussed, there are three classes of vests: Class I, Class II, and Class III. All provide a different level of visibility.

Hazard Assessment for Vest Types

All agencies involved in traffic incidents should do a hazard assessment of their work zones to determine what vest is needed for the type of work being performed. According to OSHA 1910 Subpart I App B, each agency should perform a hazard assessment with the following categories in mind:
- Impact
- Penetration

Figure 4-4 Public Safety Vest.

- Compression (rollover)
- Chemical
- Heat
- Harmful dust
- Light penetration

Some of these categories will not apply to all incidents, but all categories must be considered in any work area. In the case of a firefighter participating in firefighting operations, OSHA has allowed an exemption to allow for a variance within this section. Firefighters exposed to direct flames, fire, heat, and/or hazardous materials are exempt. A determination has been made that the NFPA-approved turnout gear is equivalent to a Class II vest. All other responders must wear a high-visibility vest.

Traffic Cones

Placement of traffic cones to funnel or channel traffic is important to traffic management. According to Section 5F-59 of the MUTCD, traffic cones should be a specific size and color. At a minimum, they should be 28 inches in height, orange, and contain two retroreflective strips (one 4 inches in width and one 6 inches in width). See Figure 4-5.

Flags and Paddles

Flags and a STOP/SLOW paddle are the acceptable tools for directing traffic (see Figure 4-6). Each should be standard equipment for traffic management. The STOP/SLOW paddle is used for direct communication with the traffic and is the preferred method of traffic control. If responders have flags, then they must use the standard signing of flaggers.

Other Warning Devices

There are other, less often used types of equipment for traffic management. One kind of channeling device is the **flare**, which is a light

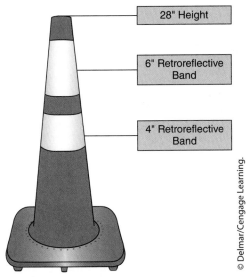

28" Height

6" Retroreflective Band

4" Retroreflective Band

© Delmar/Cengage Learning.

Figure 4-5 Traffic Cones as Specified in the MUTCD.

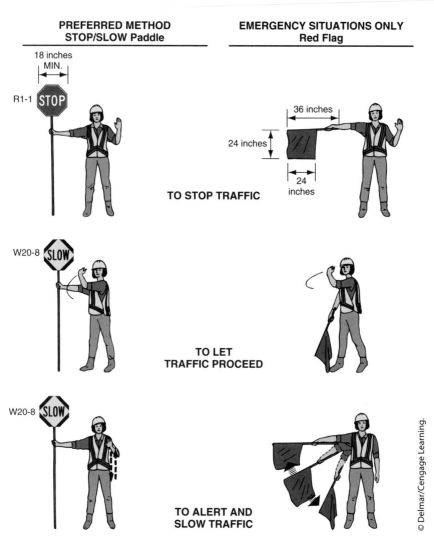

**PREFERRED METHOD
STOP/SLOW Paddle**

**EMERGENCY SITUATIONS ONLY
Red Flag**

18 inches
MIN.

R1-1

36 inches

24 inches

24 inches

TO STOP TRAFFIC

W20-8

**TO LET
TRAFFIC PROCEED**

W20-8

**TO ALERT AND
SLOW TRAFFIC**

© Delmar/Cengage Learning.

Figure 4-6 Flags and Stop/Slow Paddles.

that's bright enough to be seen from several hundred feet away. This has long been used to warn motorists of impending hazards ahead of them. Other warning devices are blinking lights. These lights can be placed on traffic cones, on the roadway, or inside traffic cones themselves.

▓▓ Flagger Emergency

A **flagger emergency** is a situation that has or is about to develop that endangers a flagger or the traffic management area. The following are examples:

- Flagger being run down
- Motor vehicle out of control
- Any situation that will endanger the crew operating at the scene
- Any unforeseen event

Audible Signal

A flagger emergency could happen at any point during a traffic management operation. **Audible signals**—warning-sound devices—must be available in addition to radio communications. A whistle or some type of horn serves this purpose, as long as the signal will alert the safety officer on scene. This procedure should be written into the SOPs. The procedure should be practiced so all responders will recognize the signal. An example of such a procedure would be sounding an apparatus horn three times in succession.

Mayday

When communicating a **mayday**—or call for help—on the radio, it should be transmitted as all maydays are in the department's SOPs. This procedure shouldn't change for roadway incidents. All messages concerning a mayday should commence with "mayday, mayday, mayday" and then end with the message that needs to be relayed to all concerned.

▓▓ SUMMARY

Having a trained flagger detail is the first defense against outside influences for a roadway incident. Flaggers should meet minimum qualifications, understand and use proper flagging equipment, and conduct operations based on safety protocol. Flaggers must also be prepared for flagger emergencies to keep all responders to the incident safe.

▓ Review Questions

1. Name three qualifications for a flagger.
2. Working on an accident scene that crests on a hill, how many flaggers will you need to keep a two-lane roadway flowing with traffic in one lane?
3. Should a flagger be placed on a curve during a highway incident? Why or why not?
4. Name 5 categories that should be considered when conducting a hazard assessment for choosing the right vest type.
5. What's a flagger emergency?

Key Terms

signage
flagger location
safety vest
flare
flagger emergency
audible signals
mayday

CHAPTER 5

Incident Organization

Learning Objectives

Upon completion of this chapter, you will be able to:

- Explain the origin of the National Incident Management System.

- Apply the incident command system to a highway incident.

- Create a typical incident management system for a highway incident.

- Develop a system interagency cooperation by using the incident management system.

Case Study

During a very dry spring, a wildland fire started in a state park. The local first responders were quickly overwhelmed by the incident because of the extremely dry conditions. The wildfire quickly grew in intensity and size within twenty four hours. The first responders quickly recognized this incident was going to need huge resources and a growing incident management team. As expected, resources grew to a reported 700 responders on-scene, as well as air operations at the height of the incident.

Once the incident management team, trained to handle an incident of this magnitude, came on scene, they quickly sized up the situation and put into place a management system that could effectively address all the concerns of the local incident commander indentified. One of these concerns was the safety of all first responders. This included the use of a main commuter road, right in the middle of the operation. The agencies responsible for this roadway quickly developed a plan that would shut down a major commuter road between two sections of the county. Once this plan was identified they needed to put the following into place:

- Design a detour route that would take traffic around the incident to make traveling safe for all.
- Place variable message signs at the detour routes.
- Place advanced warning signs well ahead of the actual detour to notify road users of this detour prior to getting there.
- Make notification to the local Traffic Management Center to use variable message signs.
- Notify the local radio stations to alert the public.

The road closure and detour were put into place for approximately three days, allowing responders to use the closed roadway without having to worry about the traveling public and curious onlookers attracted by the incident. This plan kept both the responders and the general public safer, proving that the incident command system, used correctly, ensures that responder and public safety will be addressed in a timely and effective manner.

▓▓ INTRODUCTION

Effectively managing an incident is critical to the safety and resolution of a highway accident. Responders must understand the basic application of incident management for a highway incident by using single or multiple agencies.

▓▓ The National Incident Management System

The **National Incident Management System (NIMS)** (see Figure 5-1) was developed to respond to incidents of all sizes and complexity, including incidents involving multiple agencies. It was created to handle any type of incident or event—both planned and unplanned.

NIMS COMPONENTS

- Command and Management
  - o Incident Command System
  - o Multi-Agency Coordination Systems
  - o Public Information Systems
- Preparedness
  - o Planning
  - o Training
  - o Exercises
  - o Personnel Qualifications and Certifications
  - o Equipment Acquisitions and Certification
  - o Mutual Aid
  - o Publications Management
- Resource Management
- Communications and Information Management
  - o Incident Management Communications
  - o Information Management
- Supporting Technologies
- Ongoing Management and Maintenance

© Delmar/Cengage Learning.

Figure 5-1 Components of the National Incident Management System.

The widespread use of this system can be attributed to two key elements:

1. Flexibility: The system is consistent and flexible for all agencies—at all levels—to operate. It can manage anything from a small motor vehicle accident to domestic terrorism and allows the flexibility to grow and shrink depending on what's needed at any specific time during an incident. This includes prevention, response, recovery, preparedness, and mitigation.

2. Standardization: This system provides organizational structures for multiple agencies to work with a standard set of terms and resources. In fact, resource management is relied on heavily to keep all agencies on the same level while working together.

The Basics of an Incident Command System

These key components of an **incident command system (ICS)**, such as NIMS, contribute to the operation of the system and keep management teams working professionally, with safety and accountability as the goals of each incident. The key components include: incident command, safety, operations, planning, logistics, and finance/administration (see Figure 5-2).

Span of Control

The system will always maintain a manageable **span of control**. The typical span of control under this system is the supervision of 3 to 7 people, with the optimal number being 5. If you look at the comparisons in Figure 5-3, you can see the difference in the system for a small incident (Figure 5-3A) vs. a typical large incident (Figure 5-3B).

Remember that according to MUTCD federal guidelines, responders have the responsibility to decide the magnitude of the incident and the expected duration. This fits in with the system in determining how much you'll expand or contract your incident.

Command Staff	Responsibilities
Incident Commander	• Ensures clear authority and knowledge of agency policy • Ensures incident safety • Secures an incident command post • Makes sure that incident briefing is obtained • Establishes priorities • Determines incident objectives • Establishes incident organization level and monitor its progress • Approves and implements the Incident Action Plan • Coordinates the activities of the Command and General Staff • Ensures information that is released is approved • Determines when demobilization will be done
Public Information Officer	• The PIO is responsible for interfacing with the public, media, and with other agencies with incident-related information. This person develops accurate information on the incident's size, current situation, resources committed, and other matters of internal and external purposes.
Safety Officer	• The Safety Officer monitors and tracks incident operations and advises the IC on all matters that relate to the safety of the incident, its responders, and public. The Safety Officer is responsible for the set of procedures necessary to ensure ongoing assessment of any hazardous situations. The Safety Officer has authority to stop any unsafe procedure immediately and then report them immediately.
Liaison Officer	• The Liaison Officer is the point of contact for representatives of other agency, public or private. In any command structure, representatives from cooperating or assisting agencies coordinate through the Liaison Officer.

© Delmar/Cengage Learning.

Figure 5-2 Key Positions in ICS.

(continued)

General Staff	Responsibilities

Operations Section Chief	• The Operations Chief is responsible for managing all the tactical operations at the incident. The IAP provides the necessary information and guidance for the operation. Major responsibilities include:
	  • Manage all tactical operations
	  • Assist in development of the operational part of the IAP
	  • Supervise the overall operational portion of the IAP
	  • Maintain contact with supervisors assigned
	  • Ensure all operations are safely executed
	  • Make changes when tactical operations are not working
	  • Keep close contact with Incident Commander

Planning Section Chief	• The Planning Section Chief is responsible for creation of the incident planning process by executing the following:
	  • Collect and manage all incident-related operational data
	  • Provide the IC and Operations Chief with input for creating the IAP
	  • Supervise the actual creation of the IAP
	  • Conduct and facilitate planning meetings
	  • Determine resource support in the incident
	  • Create and re-assign task forces and strike teams, as needed
	  • Create specialized positions as needed, Weather, Haz Mat
	  • Create and plan for alternative measures for contingency plans
	  • Report any significant changes in the status of the incident
	  • Create and oversee the Demobilization Plan
	  • Create specialized plans, such as traffic, medical, communications, and any other supportive action needed

Figure 5-2 *(continued)*

Logistics Section Chief	• The Logistics Chief provides all incident support functions. The following is what is expected: • Facilities • Transportation • Communications • Supplies • Equipment maintenance and fueling • Food and medical for responders • Some major responsibilities are: • Manage all logistics • Provide input to the IC concerning logistical support • Request/order resources, as needed. • Develop the communications, medical, and traffic plan • Oversee the demobilization of the logistic section
Finance/ Administration Section Chief	• The Finance/Administration Chief is responsible for the overall financial portion of the incident. Only if the agencies at hand need the support of this section will this be activated. Some responsibilities of this section are: • Manage all financial aspects of the incident • Provide cost analysis of the incident • Make sure all compensation claims are addressed • Maintain daily contacts with agency representatives • Establish and collect agency time records for the incident • Brief personnel on all incident-related finance issues

Figure 5-2 (continued)

© Delmar/Cengage Learning.

Preplanning

In the preplanning process, responders need to address resource management. Under an ICS system, agencies can have shared responsibilities, such as costs, staffing, and safety issues related to any incident.

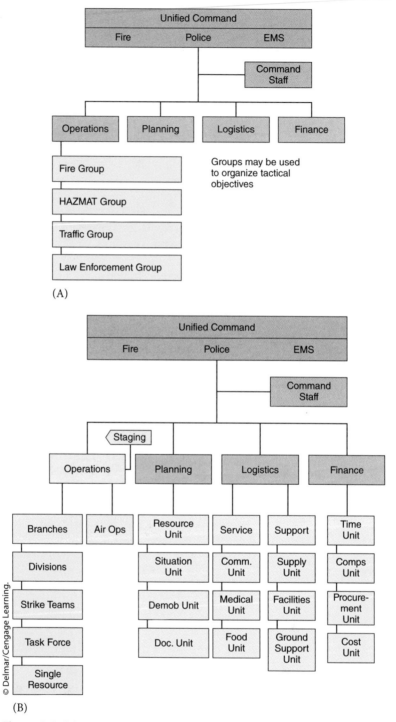

Figure 5-3 (A) System with Groups In Use (B) An expanded ICS structure.

Full Vehicle Response—This is typically when all personnel are allowed to respond to all incidents with their personal vehicles.

Partial Vehicle Response—This policy allows a limited number of personal vehicles on the highway, typically, officers and fire police only. All others will respond to the fire station and/or a predesignated staging area.

No Vehicle Response—No personal vehicles are allowed to respond. All personnel will report to the fire station and will be transported on responding apparatus or report to a predesignated staging area not on the highway.

Figure 5-4 Sample Policy for Use of Personal Vehicles.

Staging Areas

Staging areas should be designated areas for the IC to place apparatus and human resources until a decision has been made as to what mission they'll be assigned. These areas can also serve as parking areas for personal vehicles instead of having them park on the highway and add to the confusion and traffic congestion (see Figure 5-4).

Implementing ICS

The first responder on scene should initiate ICS upon arrival. Assuming that any incident in the jurisdiction has been preplanned and discussed among the local agencies, the first arriving agency assumes command of the incident, whether it's law enforcement, EMS, or fire services.

Implementing this system on arrival calls for the following actions by the IC:

- Establish command on arrival
- Assign specific duties to qualified personnel
- Decide overall objectives for the operation

- Ensure personnel assigned are maintaining their place in the system
- Confer with political and private entities that have a stake in the operation
- Obtain regular incident briefings
- Ensure open communications between your staff and those dealing with the incident
- Ensure the overall safety of the incident.

Expanding ICS or NIMS

NIMS is modular in nature and should develop from the top-down. If an IC feels the need to expand the system because the incident is growing, the system allows for this flexibility. For example, when responding to a highway incident reported to be 2 vehicles with entrapment, responders will use a small modular system of command, such as the one shown in Figure 5-5. Typically the IC, safety officer, and operations officer positions are activated. If responders arrive and see a tanker filled with hazardous materials leaking into a water supply, the IC may choose to expand the system. This type of emergency is estimated to have a 12- to 24-hour operational period, and responders may have to activate the planning section and logistics section of the command system. In this instance, the system would have grown, and the IC would have activated essential elements to manage the incident.

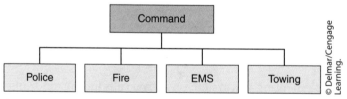

Figure 5-5 Typical Highway Incident Command Structure.

> ## Lesson Learned
>
> During a downed aircraft incident, there was a team tasked as part of the planning section. The mission of this planning section was to have a traffic plan developed for the incident. This required the development of interagency cooperation between local, county, and state agencies. The core of this mission was to reroute traffic, develop traffic control points, and staff these points 24 hours for at least 7 days. This took extensive planning. Staffing for this detail consisted of approximately 220 staff to accomplish and maintain. This sole mission during the event took a huge chunk of time to coordinate and develop. In a situation like this, the IC cannot accomplish the planning alone. This is the proven worth of the incident command system working for all involved.

Large-Scale Incidents

At **large-scale incidents**, complex incidents involving multiple agencies, this system will be extremely valuable. Consider managing a large-scale incident with over 500 responders with a single IC. Finding time to talk with the news media, coordinate resource management, track personnel, order supplies and rehab for the responders, and deciding when and where to house off-duty personnel would be overwhelming at best.

Training

All agencies—large or small—need to train and use this system in day-to-day operations. The utilization of ICS in emergency and non-emergency events will ensure ongoing training, an understanding of the system, knowledge of how the system meets the needs of everyone, and the promotion of the team concept with all involved.

To maintain the skills necessary to sustain NIMS, responders may need to implement this system during nonemergency events, such as:

- Agency exercises
- Fund-raisers
- Tabletop exercises
- Multi-agency exercises
- Training seminars

Traffic Management

For every incident—whether a highway incident or some other type of incident—traffic is always a consideration in the response. Traffic control is an integral part of planning activities. NIMS provides us with the tools needed to work on today's highways with outstanding accountability and good structuring: resource management, safety, and preplanning, among others.

Safety

The consideration of safety issues is an essential part of the NIMS system. No operation will be undertaken unless it's approved by the safety officer and won't endanger a responder and/or victim. This requirement applies to all responding agencies, including law enforcement, fire services, EMS, a towing service, and any other agency you may bring to your scene.

Inter-Agency Cooperation

Agencies responding to an incident must consider all aspects of the incident management system. As mentioned earlier, the three Cs—cooperation, communication, and coordination—are important when working with multi-agency, multi-jurisdictional incidents.

Coordination

The command system fosters this kind of thinking and allows you to overcome any issue that may arise during any operation. Without the coordination of all operational resources, responders can't effectively direct each resource to the best location. Having agencies and resources on scene without cause isn't a good management of the public's expectations.

Communication

The incident command system also emphasizes the fact that all the different agencies need good, common, and solid communications. This applies not only to command staff but also to the agencies working together—whether it's in a task force or standing side by side in the roadway. This is an essential part of any operation. One of the key elements needed to keep an incident from being disorganized and unsafe is a good, sound communications plan. Having high-tech communication systems serves no purpose unless responders are trained and the equipment is properly programmed.

Cooperation

There must be cooperation among all agencies involved. Coordination and communication will not suffice if there is not a solid foundation of cooperation.

▓▓ SUMMARY

Overall, the National Incident Management System (NIMS) is a valuable system that can and should be utilized in all incidents. It has proven to be a solid management tool to guide emergency services to command and manage any incident. This system will supply all the justification, resource management, accountability, safety, and staff needed to effectively manage any incident.

Review Questions

1. What are the three principles needed to manage a multi-agency, multi-jurisdictional incident?
2. What is the ideal span of control?
3. What position is responsible for the overall command and management of any incident under NIMS?
4. What decisions does the IC have to make upon arrival at an incident?

Key Terms

National Incident Management System (NIMS)
Incident command system (ICS)
span of control
staging areas
large-scale incidents

CHAPTER 6

Apparatus Considerations

Learning Objectives

Upon completion of this chapter, the student will be able to:

- Describe apparatus designs that create safe working environments.

- Place apparatus on the roadway for protection of the emergency responders.

- Understand lighting considerations needed to maintain a safe work area.

Case Study

A volunteer firefighter was fatally injured when a tractor-trailer struck his parked, privately owned vehicle. The volunteer firefighter had responded to a weather-related single motor-vehicle incident on an interstate highway. The vehicle was traveling eastbound when the driver lost control, drove through the median into the westbound lanes, and rolled over onto the north shoulder of the westbound lanes. Upon the volunteer firefighter's arrival to the scene, the incident commander advised the victim to position his pickup truck upstream to warn oncoming traffic of the vehicle incident in the curve. As indicated, he positioned his vehicle upstream on the north shoulder of the westbound lanes and turned on his emergency flashers and rooftop light bar. An oncoming tractor, pulling two trailers, lost control while changing lanes causing the rear trailer to swing counter-clockwise. The operator swerved several times before the rear trailer struck the victim's pickup truck, which was positioned on the north shoulder. The firefighter wasn't ejected from the vehicle but was found lying on the rear set of seats wearing a seatbelt. He was pronounced dead at the scene.

Key contributing factors identified in the investigation of this incident included the hazardous road conditions, the speed of the tractor-trailer, and the non-use of a seatbelt by the firefighter. The following recommendations were made:

- National Institute for Occupational Safety and Health (NIOSH) investigators concluded that in order to minimize the risk of similar occurrences, companies using tractor-trailers should ensure that operators drive in a manner that's compatible with weather conditions.
- NIOSH investigators also recommend that fire departments and fire service consensus standard committees consider re-evaluating current standards on seatbelt use to include their use while vehicles are parked and occupied at highway incidents.
- Although there's no evidence that the following recommendation could have specifically prevented this fatality, NIOSH investigators recommend that fire departments should re-evaluate current

policies and procedures to ensure that temporary traffic control devices are available and deployed upstream of warning vehicles.

▒ INTRODUCTION

Apparatus design, apparatus placement at the scene, and lighting considerations are important safety factors for highway incident scenes. Personnel need to understand all aspects of apparatus use and safety features in order to create a safer work area for emergency personnel.

▒ Apparatus Design

Apparatus design changes every time the fire service finds a new way of keeping personnel safer. Effective January 1, 2009, the National Fire Protection Association's *NFPA 1901 Standard for Automotive Fire Apparatus* now includes additional equipment for new fire service apparatus. This new equipment requirement specifically addresses highway safety concerns.

Specialized Safety Designs

Keeping vehicles updated with new vehicle safety designs always creates a challenge, and this is true no matter what agency operates on the roadway.

The use of fire service vehicles to warn advancing traffic is illustrated in Figure 6-1A. One department uses a tanker as their blocking or shadow vehicle. The extra compartment in the rear is used for storage of signs, cones, and vests for the responders.

Figure 6-1B is an example of what one law enforcement agency created to keep its officers safe while operations are conducted. Elevating the light package on the roof of the patrol vehicle gives oncoming traffic more advanced notice. This will slow traffic down earlier in order to keep the scene safer.

In addition, we often talk about the necessity of fire apparatus, law enforcement, and EMS vehicles as being visible. However, we also need to think about the command vehicles from each of these

(A)

(B)

© Delmar/Cengage Learning.

Figure 6-1 (A) A Recently Designed Tanker as a Buffer Vehicle with a Large Compartment in Rear for Traffic Cones. (B) Police Vehicle Suited for Heavily Traveled Roadways for Longer Awareness for Drivers. (C) A Suburban that has the Updated Striping for Safety While Out on the Roadways.

(continued)

(C)

Figure 6-1 (continued)

agencies. Figure 6-1C shows how a suburban area could be striped to warn traffic, which can be used by any agency that responds.

Equipment Specifications

The following is a list of equipment that needs to be incorporated on each new piece of apparatus:

- One traffic vest for each seating position—with each vest complying with ANSI/ISEA 207, Standard for High-Visibility Public Safety vests, and having a 5-point breakaway feature that includes two at the shoulders, two at the sides, and one at the front (refer back to Figure 2-13)
- Five fluorescent orange traffic cones not less than 28 inches in height, each equipped with a 6-inch reflective white band no more than 4 inches from the top of the cone and an additional 4-inch retro-reflective white band 2 inches below the 6-inch band (refer back to Figure 4-5)

Figure 6-2 Highway Flares that have been Prepackaged for Deployment at an Incident.

- Five illuminated warning devices, such as highway flares (Figure 6-2), unless the five fluorescent orange traffic cones have illuminating capabilities

Retro-Reflective Striping

Additionally, the following requirements are outlined in Section 5.8.3 of *NFPA 1901* under miscellaneous equipment:

- At least 50% of the rear-facing vertical services visible from the rear of the apparatus—excluding any pump panel areas not covered by a door—shall be equipped with retro-reflective striping in a chevron pattern sloping downward and away from the center-line of the vehicle at an angle of 45 degrees (see Figure 6-3).
- Each strip in the chevron shall be a single color, alternating between red and either yellow, fluorescent yellow, or fluorescent yellow-green.
- Each stripe shall be 6 inches wide.
- All retro-reflective striping materials are required to meet ASTM D 4956, Section 6.1.1.

(A)

(B)

Figure 6-3 The NFPA Standard for the Rear of Apparatus.

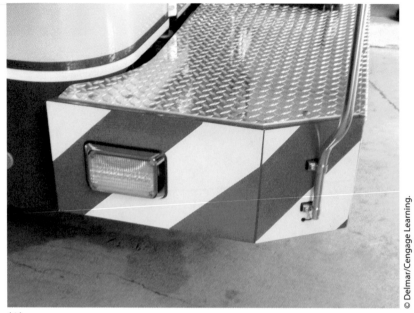

(A)

(B)

Figure 6-4 The Front Bumper of Apparatus with New Color Scheme for
Attention and Safety.

The fire service is not only adding chevrons to the rear but also in the front. Figures 6-4A and 6-4B illustrate this striping on the front and side bumper of an engine. This will help not only when the apparatus faces the front but also on the sides. This will increase the visibility all around the apparatus.

Additional Requirements

This is a significant shift toward safety of on-scene personnel. Likewise, other agencies have created special vehicles to assist with traffic control and carry the necessary equipment (see Figure 6-3).

In addition, these other agencies—including law enforcement and EMS—are starting to implement retro-reflective striping with their vehicles. Law enforcement agencies are starting to add reflective striping on the rear and lighting packages, which create greater advanced warning. EMS vehicles are now putting chevron striping on the rear of ambulances, including the inside rear doors, which will increase the reflectivity facing the rear.

(A)

Figure 6-5 A-C Various Agencies Have Created Special Vehicles to Assist with Traffic Control and Carry the Necessary Equipment.

(continued)

(B)

© Delmar/Cengage Learning.

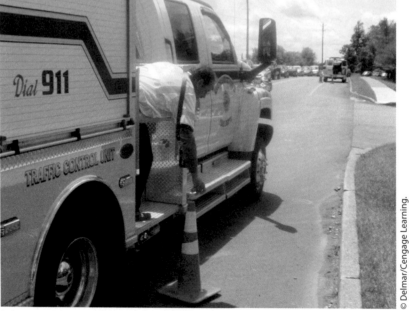

(C)

© Delmar/Cengage Learning.

Figure 6-5 (continued)

As a whole, all emergency services need to design future equipment with the safety of the responders in mind. If we look at how European response vehicles are marked, there's no mistake as to what's in the roadway. The emergency vehicle as well as all persons must be seen more than ever in order to reduce injuries and fatalities.

Placement of Apparatus

The placement of vehicles on an incident roadway is important for the safety of the responders as well as the safety of the general public. How and where each vehicle should be placed needs to be stressed in every operation. The MUTCD standard and the National Fire Protection Association (NFPA) both stress the importance of vehicle placement.

Shadow Vehicle

The MUTCD supports the use of a **shadow vehicle**—a moving vehicle that provides physical protection for workers—when operating on the

© Delmar/Cengage Learning.

Figure 6-6 A Fire Service Vehicle Properly Placed to Provide Protection for Minor Incident.

roadway. Use of this option is most common for emergency services during temporary traffic control. The placement of this vehicle is critical when trying to protect the work area from stray and/or out-of-control traffic.

NFPA 1500

NFPA 1500 Standard on Fire Department Occupational Safety and Health outlines safety standards for apparatus placement and shielding. It states: "Fire apparatus shall be positioned in a blocking position, so if it is struck it will protect members and other persons at the incident scene."[1] This adds to the safety of all personnel concerned.

The placement of apparatus will vary depending on the type of incident and roadway encountered (see Figures 6-7 through 6-10).

As mentioned previously, the following factors need to be considered:

- Type of roadway
- Curvature of the roadway
- Speed
- Closing of lanes
- Signage

Placement Guidelines

The recommended apparatus placement may require additional highway responses from other agencies. This would depend on the duration of the incident and resources from the responding agencies. At a minimum, a **blocker vehicle**—the initial on-scene emergency vehicle—should be placed at the transition area for protection of the queue length. This will create added protection and warning for responders on the scene.

Some apparatus placement guidelines emergency services should follow include:

- Apparatus should be angled toward the traffic lanes.
- All personnel should use protective vests.
- Apparatus should carry traffic cones for setting up the transition area and traffic queue.
- The working apparatus should be angled to protect the work area when a blocker vehicle is used. This will protect the driver and workers on scene.
- Appropriate vehicle lighting should be used.
- Any additional vehicles, such as tow trucks or ambulances, should be placed at the end of the queue for protection and ease of departure.

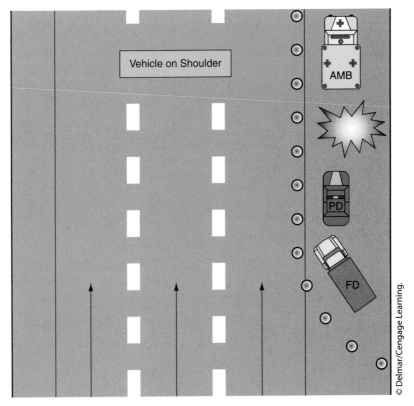

Figure 6-7 Shoulder Incident.

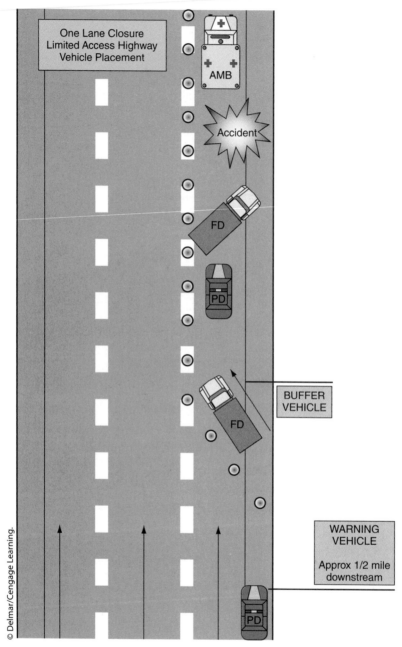

Figure 6-8 Incident Requiring Lane Closure.

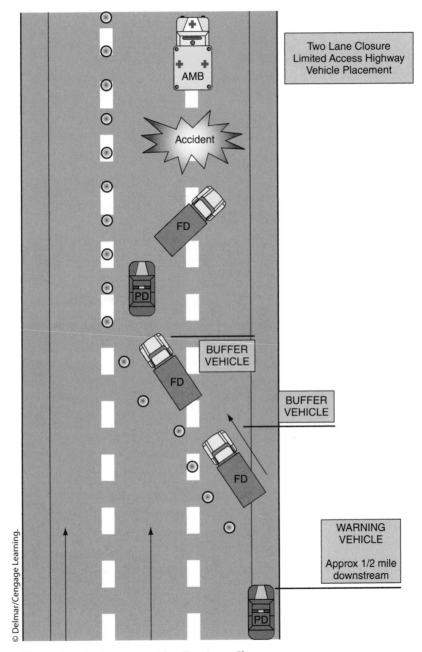

Figure 6-9 Incident Requiring Two Lane Closures.

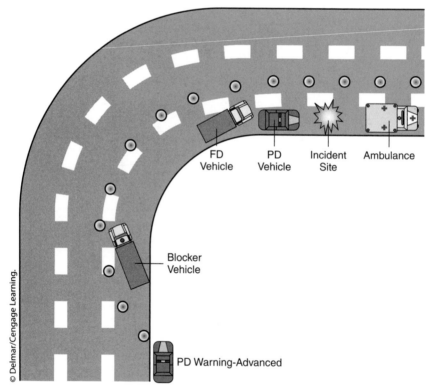

© Delmar/Cengage Learning.

Figure 6-10 Incident Requiring Lane Closure with Curve.

▓▓ Lighting Considerations

Lighting for an accident scene is important for scene safety. Looking at emergency lighting in general, it's been designed for use en route to an emergency scene. It has also been used for scene safety. However, apparatus are now required to have one yellow warning light on the rear of apparatus, similar to the yellow warning lights used by road crews (see Figure 6-11).

Figure 6-11 A Properly Outfitted Fire Service Vehicle with Clear Directional Arrow.

Examples of Vehicle Lighting

Some examples of vehicle lighting are:

- Arrow sticks on the rear and front of apparatus used to direct traffic
- Arrow boards for traffic flow direction
- Sufficiently powered lights

Headlights and Emergency Warning Lights

The human eye requires time to recover from lighting, such as headlights. In fact, while driving, drivers should look away from headlights due to the time it takes to recover from being exposed to the intense light. If you're traveling at a high speed, looking directly at a headlight may render you unable to see anything for approximately 6 seconds, which could put emergency workers in danger.

To reduce the element of danger, we should try and accomplish the following tasks:

1. Reduce glare from lights by shutting off headlights and emergency lighting
2. Have scene lighting aimed away from the flow of traffic
3. Raise lighting so it's pointed downward on the incident, improving sight for responders and keeping lighting from blinding the flow of traffic

▓ SUMMARY

The use of vehicles and vehicle lighting is one of the most important safety factors at a highway incident. The use of blocker vehicles to protect you and other agency personnel can make or break a safe operation. On-scene responders need to have confidence that command will set up the scene with their safety in mind.

Review Questions

1. What is a shadow vehicle?
2. Should the emergency warning lights used en route to an emergency be used to light a highway incident? Why or why not?
3. When designing apparatus, what safety features should be added to the rear of the vehicle?

Key Terms

shadow vehicle
blocker vehicle

Endnote

1. *NFPA 1500 Standard on Fire Department Occupational Safety and Health*, Chapter 8, 2007 edition.

CHAPTER 7

Preplanning Considerations

Learning Objectives

Upon completion of this chapter, the student will be able to:

● Define traffic management area.

● Conduct a preplanning meeting.

● Explain the need for preplanning meetings among all emergency responder organizations.

● Plan and utilize detours.

● Set up a safe and effective landing zone.

Case Study

First responders were called to a commercial-truck fire along an interstate on a breezy fall afternoon. Upon arrival responders found a tanker truck with its brakes on fire. The fire caused the tanker to leak acid onto the roadway. This leak resulted in acid draining into a local water shed along the interstate. Upon discovery of this incident, the incident commander immediately shut the roadway down and called the local hazardous materials team to mitigate the leak on the tanker.

Traffic became congested at least ten miles from the incident. The first responders not only had to deal with the hazardous material spill, they also needed to plan for traffic management. They needed to reroute traffic around this incident which entailed contacting local police resources to handle local traffic. They also needed to contact the Department of Transportation and the local Traffic Management center to create a detour route in the immediate area and design a detour outside the immediate area to prevent traffic from coming in. In addition, they needed variable message boards and local radio stations to notify the public of this incident.

The incident commander set up an incident command team to deal with the following circumstances:

- Control the leak from the tanker.
- Evacuate residents around the incident because the chemical was creating a cloud.
- Shut down the interstate traffic and reroute traffic.
- Make certain that proper resources were called to handle the incident.
- Notify the public of the evacuations and rerouting of traffic.

The incident took some time to plan and took numerous resources to handle. Preplanning such incidents and identifying local resources helps keep the issues relating to such an incident to a minimum.

▨ INTRODUCTION

When agencies conduct **preplanning meetings** together, they establish unified objectives and determine what will be required from each individual agency during any given emergency response. Preplanning can be effective no matter how large or small the incident.

▨ Preplanning with Other Emergency Response Organizations

All organizations that may potentially respond to an incident need to consider that traffic will need to be managed. Whether at a motor vehicle accident or a full-fledged wildland fire, traffic is an important factor in planning to managing the aftermath of an incident.

LESSON LEARNED

Recently, a team was tasked with working in the planning section of a considerably sized wildfire. This fire covered approximately 4,500 acres of forest land. One planning consideration was traffic management.

The team needed to plan on some important issues within and on the perimeter of this incident:

- Should a main commuter traffic route be closed?
- If so, how would traffic be rerouted?
- What signage—if any—is needed?
- How would the public be informed?
- Would this create a political issue for the team?

After reviewing its options, the team did close the route for the safety of the operation and public. Planning took place in conjunction with the local law enforcement, political heads in all communities, the incident commander, and the operations

Figure 7-1 Variable Message Board - Portable.

chief. This plan was incorporated into the overall incident action plan, and the other responders were advised of these actions. The public was notified via newspapers and public information notices through the traffic management center and radio stations. Variable message signs were acquired from the Department of Transportation (DOT) to be deployed before and around the detours for ease of travel (see Figure 7-1).

Preplanning Considerations

The following organizations will need to be a part of preplanning meetings:

- All emergency services organizations
- Local and state department of transportation
- Hazmat teams
- Communications centers

- Traffic management centers
- Department of Environmental Conservation
- Towing companies
- Utility companies
- Air, rail, or ground transportation organizations

When conducting a preplanning meeting, specific considerations and/or objectives should be listed ahead of time. The following are some possible issues for consideration:

- Fire service actions
- Law enforcement actions
- EMS actions
- Placement of vehicles at the traffic management area
- Staging of vehicles and organizations
- Actions each will conduct
- Unified command structure
- Personal protective equipment
- Traffic detour routes and plans
- Specialized team support
- DOT support for long-term incidents

Having preplanning meetings with agencies at least once a year will keep everyone up to date on key items, such as changes in policies and officers, agency goals, and reviewing previous issues that may have presented themselves since the last session.

Conducting Preplanning Meetings

A preplanning meeting provides an opportunity to solve and discuss issues that are common to all agencies and to discuss the mission of each agency. Each agency member represented at a preplanning meeting should have extensive knowledge of his or her agency's objectives. The representative should also be able to make decisions on behalf of his or her agency.

Some of the issues addressed at a preplanning meeting would be:

- Agency goals
- Agency concerns

- Command structure at various incident types
- Preplanning different incidents
- Policies created and why
- Detour routes
- Response routes
- Any considerations with safety, roadway issues, traffic management, scene safety, and media control

All participants attending these meetings need to have the three Cs in mind:

- Coordination
- Cooperation
- Communication

According to the MUTCD, emergency service agencies need to help reduce traffic flow in times of unusual incidents. Consulting with indirect agencies—such as utility companies, schools, railroads, etc.—will reduce the chance of secondary crashes by establishing alternative travel routes. Public radio, dispatch centers (which can contact schools), utility companies, and the railroad will help reduce traffic flow at an incident site. This will keep on-scene responders safer with less traffic flow going through the incident site.

Preplanning Detour Considerations

During the initial size-up of the situation, responders need to consider the use of **detour routes,** which are alternative paths for the flow of traffic. Having traffic sit at a standstill has at times proven to be detrimental to an incident scene as well as motorists (see Figure 7-2). Secondary accidents have occurred during traffic tie-ups, and these secondary incidents could also interfere with the response to the original incident.

Detour Route Considerations

Some considerations when planning detour routes should be:

- Is the detour simple to follow?
- Will the detour accommodate oversized vehicles?

Figure 7-2 Traffic Backup During an Incident.

- Will commercial traffic be able to clear bridges?
- Do you have enough manpower to coordinate the detour route planned?

Detour Signage

Signage is an important tool for detour routes (see Figure 7-3). Signage can enhance the traffic management and be used in place of personnel needed for more essential duties.

Knowing where to obtain signage and other equipment is part of the preplanning process. Knowing where to obtain variable message signs is vital to a long-term incident, and making use of prepositioned message boards can be essential. Working with the DOT to use these signs is an option to be considered and can be arranged during preplanning meetings. The availability of changing or putting these signs into operation would then be only a phone call away.

Figure 7-3 Detour Sign at Exit.

Preplanning Traffic Management

Signage and traffic diversions are an essential part of a **traffic management plan,** which is the plan for mitigating any and all anticipated traffic problems. The use of certain signs at the beginning of an incident is as important as the use of extended term signage. **NFPA 1500** recommends signage for initial operations. Some agencies have put signs on the first response vehicles to slow traffic down prior to coming upon an incident. For example, a sign placed well before an incident that reads "Emergency Scene Ahead" (see Figure 7-4) will help responders slow traffic down or make motorists aware of an impending traffic pattern change.

The use of cones placed on the blocking vehicle to channel traffic within a lane of traffic will help make motorists aware and channel them away from the incident. Agencies that can assist with traffic detours need to be included in the planning process. Every agency—after attending the planning meeting—may have to change policies or procedures to incorporate the response established during preplanning.

Figure 7-4 Emergency Scene Ahead.

Traffic Management Center (TMC)

As previously discussed, the traffic management center (TMC) has all the operations for traffic management at its fingertips and can remotely send messages to temporary message boards, permanent message boards, and close and open lanes right from the TMC dispatch floor. Agencies should utilize this valuable resource in every preplan for an efficient traffic management zone (see Figure 7-5).

Department of Transportation

Traditionally, a **Department of Transportation (DOT)** unit—a governmental agency that oversees all forms of transportation—isn't an emergency response agency. If fire departments want the DOT to respond to emergency scenes, the DOT should be asked to attend the

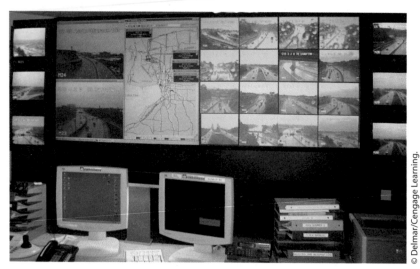

Figure 7-5 Traffic Management Control Center.

planning meeting. The DOT can then learn what the expectations are and be able to plan how to carry out its part of the incident response. Calling the DOT only when an incident occurs is ineffective and may have unexpected results.

DOT agencies are a valuable resource for safe and effective traffic management. Although not traditionally considered a response agency, the DOT is now formulating plans to assist in a more efficient manner with the emergency services. Some agencies have prepared vehicles, such as trailers, with equipment to quickly respond to a call to relieve the emergency services of traffic management so they can concentrate more on the mission at hand: saving lives.

Preplanning Helicopter Operations

Helicopter operations are essential when responding to highway incidents and even non-highway incidents. Preplanning and having operating guidelines for helicopters will keep all personnel safe.

Figure 7-6 Some Agencies are Putting Together Quick Response Trailers to Assist with Managing the Scene and Meeting the MUTCD Standard.

Landing Zone

The following items need to be addressed for safe operations for the area when the helicopter will land, or the **landing zone** (LZ):

- Placement of the landing zone
- Security of the landing zone
- Operation of the landing zone
- Establishing proper communications
- Safe area of operations

The LZ placement is vital to safety operations and patient care. Putting a helicopter on the highway may not be the safest place. Responders must determine if highway placement will help the operation or make the area unsafe.

Preplanning landing zones will need to be a priority. Working with the pilots, law enforcement, EMS, and fire services will help to determine the best options. Placing the LZ off the roadway will also be a consideration when looking for sites. Some DOT officials have been considering placing LZs off the highway, which would provide better scene safety, personnel safety, patient care efficiency, and quick clearance of the roadway.

Another consideration of LZs is proper setup (see Figure 7-7). Having an area cleared for a LZ isn't the only consideration. Other considerations include:

- 100 feet by 100 feet minimum ground clearance
- Firm and level ground
- No overhead obstructions, such as trees, power lines, and towers (see Figure 7-8)
- No loose debris
- Absence of blinding light sources, especially during night landings

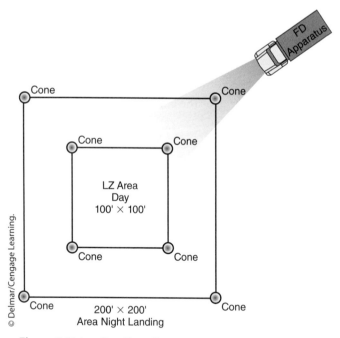

Figure 7-7 Landing Zone Setup.

Figure 7-8 Helicopter with Power Lines and Towers.

Security and Communication

Landing a helicopter tends to draw the attention of curious onlookers and emergency services personnel alike. Because of this, responders need to properly secure the area and ensure a safe approach (see Figure 7-9).

Figure 7-9 Safe Approach of Aircraft.

Proper communication needs to be established prior to the operation. Some agencies have predetermined radio frequencies dedicated for such operations. Usually, the helicopter—given the proper frequency and code—can tune into any radio on scene.

SUMMARY

Overall, the preplanning of any operation is essential to the success of that operation, whether it's an emergency or a planned event. Members of the emergency services need to meet and discuss goals and objectives for operating at highway incidents. All agencies—emergency and non-emergency—need to work together to serve the public. Whether it entails changing some standard procedures or retraining personnel, this needs to be a priority to help save time, lives, and money.

Review Questions

1. What's the minimum size of a landing zone for helicopter landings?
2. What's the value of a planning meeting with multiple agencies that respond to a highway traffic incident?
3. Name 4 agencies that should be included in traffic management preplanning meetings.

Key Terms

preplanning meetings
detour routes
traffic management plan
Department of Transportation
landing zone

Appendix A

Glossary

A

activity area: Where emergency service work actually takes place. This area must be designated by the incident commander and protected for safe operations.

advanced warning area: Where motorists are first informed about the upcoming traffic pattern in your emergency work zone.

American National Standards Institute (ANSI): A private, nonprofit organization promoting a voluntary consensus of standards for products to ensure safety and health.

audible signals: Warning-sound devices that must be available in addition to radio communications. A whistle or some type of horn serves this purpose, as long as the signal will alert the safety officer on scene.

B

blocker vehicle: This initial on-scene emergency vehicle should be placed at the transition area for protection of the queue length. This will create added protection and warning for responders on the scene.

C

climbing lanes: Auxiliary lanes provided for slow-moving vehicles ascending steep grades. They may be used along all types of roadways.

collector roads: They collect and distribute traffic while providing access to abutting properties.

D

Department of Transportation (DOT): A governmental agency that oversees all forms of transportation.

detour routes: Alternative paths for the flow of traffic.

F

flagger emergency: A situation that has or is about to develop that endangers a flagger or the traffic management area.

flagger location: The placement of the flagger.

flare: A light that's bright enough to be seen from several hundred feet away.

H

hot zone: The work zone of the traffic management area.

I

incident command system (ICS): Contributes to the operation and keeps management teams working professionally, with safety and accountability as the goals of each incident.

incident within an incident: Even with an extra person assigned to the traffic backup, the traffic backup can at times become an incident itself.

intermediate incident: An accident that will likely take between 30 minutes and 2 hours to deal with. Intermediate incidents are the most typical incidents to which a fire service responds. These incidents are usually car fires or one- or two-car motor vehicle accidents.

International Safety Equipment Association (ISEA): A trade association whose member companies promote the protection of health and safety for workers.

interstates: Limited-access highways that are generally inter-regional, high-speed, high-volume, and divided roadways with complete controlled access.

L

landing zone: The placement of this is vital to safety operations and patient care.

large-scale incidents: Complex incidents involving multiple agencies.

line of sight: The direction from which traffic comes is important when considering where to place cones and set up signs.

M

major incident: An accident that takes more than 2 hours to deal with. This incident requires substantial time and resources to bring under control. Examples would be a multiple vehicle incident, a hazardous materials incident, a bridge collapse, or any other incident that would require extended operations.

Manual on Uniform Traffic Control Devices (MUTCD): Defines the standards used by road managers nationwide to install and maintain traffic control devices on all public streets, highways, bikeways, and private roads open to public traffic.

mayday: A call for help that should commence with "mayday, mayday, mayday" and then end with the message that needs to be relayed to all concerned.

minor incident: An accident that will take less than 30 minutes to remove the road user from the lanes of travel. An example would be responding to an emergency that requires EMS but where the vehicle is off the roadway.

move-over law: This law requires motorists traveling on multi-lane roadways to—when practical—merge away from a vehicle working on the side of the highway to provide an empty travel lane of safety for workers.

N

National Fire Protection Association (NFPA): A nonprofit organization that develops, publishes, and disseminates codes and standards intended to minimize the possibility and effects of fire and other risks.

National Incident Management System (NIMS): Developed to respond to incidents of all sizes and complexity, including incidents involving multiple agencies. It was created to handle any type of incident or event—both planned and unplanned.

National Institute for Occupational Safety and Health (NIOSH): The governmental agency responsible for conducting research and making recommendations for the prevention of work-related illnesses and injuries.

O

operations officer: This person should monitor what's happening within the hot zone. This position ensures that responders are following good, sound safety practices within the actual operation.

P

portable variable message signs: These display changeable electronic messages and are available to incorporate on your vehicle and apparatus.

preplanning meetings: When agencies establish unified objectives and determine what will be required from each individual agency during any given emergency response. Preplanning can be effective no matter how large or small the incident.

Q

queue length: Backed-up traffic.

quick clearance policy: This is rapidly becoming the basis for reducing traffic incident injuries and fatalities among emergency responders. A quick clearance policy is one that promotes fast cleanup of the traffic incident to reduce the traffic backup as soon as possible.

R

ramps: Are turning roadways that connect two or more sections of an interstate.

recovery process: The means by which a problem is resolved. Sometimes, this step can be longer than the original incident.

rehabilitation: Getting the traffic flow moving again.

S

safety officer: The person responsible for all responders at an incident.

safety vest: One of the primary safety items issued to a flagger. According to the MUTCD, flaggers must wear a vest to protect themselves from traffic. The reason for wearing a safety vest is to be seen.

secondary crash: Commonly occurs when the initial crash causes traffic within the lanes to back up and create congestion.

shadow vehicle: A moving vehicle that provides physical protection for workers when they're operating on the roadway. Use of this option is most common for emergency services during temporary traffic control. The placement of this vehicle is critical when trying to protect the work area from stray and/or out-of-control traffic.

signage: Depending on your response area, signs may be needed to warn or direct motorists. Keeping the motoring public informed and aware of your presence is important to the safety of the flagger and the overall scene.

span of control: The supervision of 3 to 7 people, with the optimal number being 5.

staging areas: Should be designated areas for the placement of apparatus and human resources until a decision has been made as to what mission they'll be assigned. These areas can also serve as parking areas for personal vehicles instead of having them park on the highway and add to the confusion and traffic congestion.

T

temporary traffic control zones: Used to modify the flow of traffic near a highway incident should be implemented by using signs, signals, or markings that conform to the standards of the MUTCD.

temporary traffic control (TTC): The needs and control

of all road users through this zone are an essential part of highway construction, utility work, maintenance operations, and the management of traffic incidents.

termination area: The end of the area in which the emergency services operate. This area is where traffic returns to its normal path of travel and speed.

traffic incident management area: An area of a highway where temporary traffic controls are installed after a roadway incident.

traffic incident: An emergency road user incident, natural disaster, or another unplanned event that affects or impedes the normal flow of traffic.

traffic management centers: The centralized location where traffic-related data are collected and from which all traffic-related information can be disseminated.

traffic management plan: The plan for mitigating any and all anticipated traffic problems.

transition area: Where traffic is diverted from the normal path of travel. This area tells the motorist that an action is required because of a lane shutdown, a lane diminished, or another impending action.

Appendix B

MUTCD, 2009 Edition, Part 6: Temporary Traffic Control

CHAPTER 6I. CONTROL OF TRAFFIC THROUGH TRAFFIC INCIDENT MANAGEMENT AREAS

Section 6I.01 General

Support:

01   The National Incident Management System (NIMS) requires the use of the Incident Command System (ICS) at traffic incident management scenes.

02   A traffic incident is an emergency road user occurrence, a natural disaster, or other unplanned event that affects or impedes the normal flow of traffic.

03   A traffic incident management area is an area of a highway where temporary traffic controls are installed, as authorized by a public authority or the official having jurisdiction of the roadway, in response to a road user incident, natural disaster, hazardous material spill, or other unplanned incident. It is a type of TTC zone and extends from the first warning device (such as a sign, light, or cone) to the last TTC device or to a point where vehicles return to the original lane alignment and are clear of the incident.

04   Traffic incidents can be divided into three general classes of duration, each of which has unique traffic control characteristics and needs. These classes are:

  A. Major—expected duration of more than 2 hours,
  B. Intermediate—expected duration of 30 minutes to 2 hours, and
  C. Minor—expected duration under 30 minutes.

05   The primary functions of TTC at a traffic incident management area are to inform road users of the incident and to provide guidance information on the path to follow through the incident area. Alerting road users and establishing a well defined path to guide road users through the incident area will serve to protect the incident responders and those involved in working at the incident scene and will aid in moving road users expeditiously past or around the traffic incident, will reduce the likelihood of secondary traffic crashes, and will preclude unnecessary use of the surrounding local road system. Examples include a stalled vehicle blocking a lane, a traffic crash blocking the traveled way, a hazardous material spill along a highway, and natural disasters such as floods and severe storm damage.

Guidance:

06   *In order to reduce response time for traffic incidents, highway agencies, appropriate public safety agencies (law enforcement, fire and rescue, emergency communications, emergency medical, and other emergency management), and private sector responders (towing and recovery and hazardous materials contractors) should mutually plan for occurrences of traffic incidents along the major and heavily traveled highway and street system.*

07   *On-scene responder organizations should train their personnel in TTC practices for accomplishing their tasks in and near traffic and in the requirements for traffic incident management contained in this Manual. On-scene responders should take measures to move the incident off the traveled roadway or to provide for appropriate warning. All on-scene responders and news media personnel should constantly be aware of their visibility to oncoming traffic and wear high-visibility apparel.*

08   *Emergency vehicles should be safe-positioned (see definition in Section 1A.13) such that traffic flow through the incident scene is optimized. All emergency vehicles that subsequently arrive should be positioned in a manner that does not interfere with the established temporary traffic flow.*

09   *Responders arriving at a traffic incident should estimate the magnitude of the traffic incident, the expected time duration of the traffic incident, and the expected vehicle queue length, and then should set up the appropriate temporary traffic controls for these estimates.*

Option:

10   Warning and guide signs used for TTC traffic incident management situations may have a black legend and border on a fluorescent pink background (see Figure 6I-1).

Support:

11   While some traffic incidents might be anticipated and planned for, emergencies and disasters might pose more severe and unpredictable problems. The ability to quickly install proper temporary traffic controls might greatly reduce the effects of an incident, such as secondary crashes or excessive traffic delays. An essential part of fire, rescue, spill clean-up, highway agency, and enforcement activities is the proper control of road users through the traffic incident management area in order to protect responders, victims, and other personnel at the site. These operations might need corroborating legislative authority for the implementation and enforcement of appropriate road user regulations, parking controls, and speed zoning. It is desirable for these statutes to provide sufficient flexibility in the authority for, and implementation of, TTC to respond to the needs of changing conditions found in traffic incident management areas.

Figure 6I-1. Examples of Traffic Incident Management Area Signs

| W3-4 | W4-2 | W9-3 | E5-2a |

| M4-8a | M4-9 | M4-10 |

Option:

12   For traffic incidents, particularly those of an emergency nature, TTC devices on hand may be used for the initial response as long as they do not themselves create unnecessary additional hazards.

Section 6I.02 Major Traffic Incidents

Support:

01   Major traffic incidents are typically traffic incidents involving hazardous materials, fatal traffic crashes involving numerous vehicles, and other natural or man-made disasters. These traffic incidents typically involve closing all or part of a roadway facility for a period exceeding 2 hours.

Guidance:

02   *If the traffic incident is anticipated to last more than 24 hours, applicable procedures and devices set forth in other Chapters of Part 6 should be used.*

Support:

03   A road closure can be caused by a traffic incident such as a road user crash that blocks the traveled way. Road users are usually diverted through lane shifts or detoured around the traffic incident and back to the original roadway. A combination of traffic engineering and enforcement preparations is needed to determine the detour route, and to install, maintain or operate, and then to remove the necessary traffic control devices when the detour is terminated. Large trucks are a significant concern in such a detour, especially when detouring them from a controlled-access roadway onto local or arterial streets.

04   During traffic incidents, large trucks might need to follow a route separate from that of automobiles because of bridge, weight, clearance, or geometric restrictions. Also, vehicles carrying hazardous material might need to follow a different route from other vehicles.

05   Some traffic incidents such as hazardous material spills might require closure of an entire highway. Through road users must have adequate guidance around the traffic incident. Maintaining good public relations is desirable. The cooperation of the news media in publicizing the existence of, and reasons for, traffic incident management areas and their TTC can be of great assistance in keeping road users and the general public well informed.

06   The establishment, maintenance, and prompt removal of lane diversions can be effectively managed by interagency planning that includes representatives of highway and public safety agencies.

Guidance:

07   *All traffic control devices needed to set up the TTC at a traffic incident should be available so that they can be readily deployed for all major traffic incidents. The TTC should include the proper traffic diversions, tapered lane closures, and upstream warning devices to alert traffic approaching the queue and to encourage early diversion to an appropriate alternative route.*

08   *Attention should be paid to the upstream end of the traffic queue such that warning is given to road users approaching the back of the queue.*

09   *If manual traffic control is needed, it should be provided by qualified flaggers or uniformed law enforcement officers.*

Option:

10   If flaggers are used to provide traffic control for an incident management situation, the flaggers may use appropriate traffic control devices that are readily available or that can be brought to the traffic incident scene on short notice.

Guidance:

11   *When light sticks or flares are used to establish the initial traffic control at incident scenes, channelizing devices (see Section 6F.63) should be installed as soon thereafter as practical.*

Option:

12   The light sticks or flares may remain in place if they are being used to supplement the channelizing devices.

Guidance:

13   *The light sticks, flares, and channelizing devices should be removed after the incident is terminated.*

Section 6I.03 Intermediate Traffic Incidents

Support:

01   Intermediate traffic incidents typically affect travel lanes for a time period of 30 minutes to 2 hours, and usually require traffic control on the scene to divert road users past the blockage. Full roadway closures might be needed for short periods during traffic incident clearance to allow traffic incident responders to accomplish their tasks.

02   The establishment, maintenance, and prompt removal of lane diversions can be effectively managed by interagency planning that includes representatives of highway and public safety agencies.

Guidance:

03   *All traffic control devices needed to set up the TTC at a traffic incident should be available so that they can be readily deployed for intermediate traffic incidents. The TTC should include the proper traffic diversions, tapered lane closures, and upstream warning devices to alert traffic approaching the queue and to encourage early diversion to an appropriate alternative route.*

04   *Attention should be paid to the upstream end of the traffic queue such that warning is given to road users approaching the back of the queue.*

05   *If manual traffic control is needed, it should be provided by qualified flaggers or uniformed law enforcement officers.*

Option:

06   If flaggers are used to provide traffic control for an incident management situation, the flaggers may use appropriate traffic control devices that are readily available or that can be brought to the traffic incident scene on short notice.

Guidance:

07   *When light sticks or flares are used to establish the initial traffic control at incident scenes, channelizing devices (see Section 6F.63) should be installed as soon thereafter as practical.*

Option:

08   The light sticks or flares may remain in place if they are being used to supplement the channelizing devices.

Guidance:

09   *The light sticks, flares, and channelizing devices should be removed after the incident is terminated.*

Section 6I.04 Minor Traffic Incidents

Support:

01   Minor traffic incidents are typically disabled vehicles and minor crashes that result in lane closures of less than 30 minutes. On-scene responders are typically law enforcement and towing companies, and occasionally highway agency service patrol vehicles.

02   Diversion of traffic into other lanes is often not needed or is needed only briefly. It is not generally possible or practical to set up a lane closure with traffic control devices for a minor traffic incident. Traffic control is the responsibility of on-scene responders.

Guidance:

03   *When a minor traffic incident blocks a travel lane, it should be removed from that lane to the shoulder as quickly as possible.*

Section 6I.05 Use of Emergency-Vehicle Lighting

Support:

01   The use of emergency-vehicle lighting (such as high-intensity rotating, flashing, oscillating, or strobe lights) is essential, especially in the initial stages of a traffic incident, for the safety of emergency responders and persons involved in the traffic incident, as well as road users approaching the traffic incident. Emergency-vehicle lighting, however, provides warning only and provides no effective traffic control. The use of too many lights at an incident scene can be distracting and can create confusion for approaching road users, especially at night. Road users approaching the traffic incident from the opposite direction on a divided facility are often distracted by emergency-vehicle lighting and slow their vehicles to look at the traffic incident posing a hazard to themselves and others traveling in their direction.

02   The use of emergency-vehicle lighting can be reduced if good traffic control has been established at a traffic incident scene. This is especially true for major traffic incidents that might involve a number of emergency vehicles. If good traffic control is established through placement of advanced warning signs and traffic control devices to divert or detour traffic, then public safety agencies can perform their tasks on scene with minimal emergency-vehicle lighting.

Guidance:

03   *Public safety agencies should examine their policies on the use of emergency-vehicle lighting, especially after a traffic incident scene is secured, with the intent of reducing the use of this lighting as much as possible while not endangering those at the scene. Special consideration should be given to reducing or extinguishing forward facing emergency-vehicle lighting, especially on divided roadways, to reduce distractions to oncoming road users.*

04   *Because the glare from floodlights or vehicle headlights can impair the nighttime vision of approaching road users, any floodlights or vehicle headlights that are not needed for illumination, or to provide notice to other road users of an incident response vehicle being in an unexpected location, should be turned off at night.*

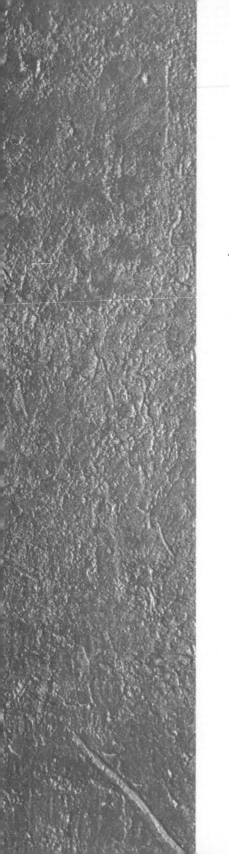

Appendix C

Highway Safety
Incident Checklist

Highway Safety Incident Checklist

Location: _____ Date: _____ Incident #: _____

Incident Commander: _____ Safety Officer(s): _____

Apparatus Assigned: _____

Incident Type: _____

Estimated on Scene Time: ☐ Less than 15 minutes   ☐ 15 minutes– 2 hours   ☐ Over 2 hours

Time of Incident: _____ Hrs   Type of Roadway: _____ Speed Limit: _____MPH

Weather Conditions: _____ Lighting Conditions: _____

Traffic Conditions on highway: ☐ Light   ☐ Moderate   ☐ Heavy

Traffic Control Actions

Blocker Vehicle in position	Communications established
Safety vests worn by responders	Termination area established (Downstream)
Advanced warning signs deployed	Flaggers stations established
Traffic cones deployed	Emergency scene warning established
Scene lighting established (not facing traffic)	Safety officer appointed
Law enforcement notified	Traffic control points established (2 hours or more)
Staging area established	Detour routes established (2 hours or more)
Work area established	DOT notified (2 hours or more)
Hazard assessment conducted	

Setup Equipment

Traffic cones
Emergency Scene Ahead Signs
Flagger Ahead Signs
Stop/Slow Paddles
Portable Radios
Safety Vests
Road Flares

Appendix D

Incident Command Forms

INCIDENT BRIEFING	1. Incident Name		2. Date	3. Time

4. Map Sketch

5. Current Organization

INCIDENT COMMANDER	
DEPUTY	
SAFETY OFFICER	
LIAISON OFFICER	
INFORMATION OFFICER	
OPERATIONS SECTION CHIEF	
BRANCH I DIRECTOR	
DIVISION SUP	
DIVISION SUP	
DIVISION SUP	
GROUP SUPERVISOR	
BRANCH II DIRECTOR	
DIVISION / GROUP SUPERVISOR	
DIVISION / GROUP SUPERVISOR	
DIVISION / GROUP SUPERVISOR	
DIVISION / GROUP SUPERVISOR	
BRANCH III DIRECTOR	
DIVISION / GROUP SUPERVISOR	
DIVISION / GROUP SUPERVISOR	
DIVISION / GROUP SUPERVISOR	
DIVISION / GROUP SUPERVISOR	
AIR OPERATIONS BRANCH DIRECTOR	
AIR SUPPORT SUPERVISOR	
AIR ATTACK SUPERVISOR	
AIR TANKER COORD. SUPERVISOR	
HELICOPTER COORD. SUPERVISOR	
PLANNING SECTION CHIEF	
LOGISTICS SECTION CHIEF	
FINANCE SECTION CHIEF	

PAGE 1 OF 2	6. Prepared by (Name and Position)

ICS 201-1

7. Resources Summary				
Resources Ordered	Resource Identification	ETA	On Scene	Location/Assignment

7. Summary of Current Actions

PAGE 2 OF 2

ICS 201-2

INCIDENT OBJECTIVES	1. Incident Name	2. Date	3. Time

4. Operational Period

5. General Control Objectives for the Incident (Include Alternatives)

6. Weather Forecast for Period

7. General Safety Message

8.		Attachments (Mark if Attached)		
☐ Organizational List - 203	☐ Medical Plan - 206		☐ (Other)	
☐ Division Assignment Lists - 204	☐ Incident Map		☐	
☐ Communications Plan - 205	☐ Traffic Plan		☐	
9. Prepared by (Planning Section Chief)		10. Approved by (Incident Commander)		

ICS 202

ORGANIZATION ASSIGNMENT LIST		1. Incident Name	
2. Date	3. Time	4. Operational Period	

Position	Name	Position	Name
5. Incident Commander and Staff		9. Operations Section	
Incident Commander		Chief	
Deputy		Deputy	
Safety Officer		a. Branch I - Divisions / Groups	
Liaison Officer		Branch Director	
Information Officer		Deputy	
6. Agency Representative		Division / Group	
Agency	Name	Division / Group	
		Division / Group	
		Division / Group	
		Division / Group	
		b. Branch II - Divisions / Groups	
		Branch Director	
		Deputy	
		Division / Group	
		Division / Group	
		Division / Group	
		Division / Group	
		Division / Group	
		Division / Group	
		c. Branch III - Divisions / Groups	
		Branch Director	
		Deputy	
		Division / Group	
		Division / Group	
		Division / Group	
7. Planning Section		Division / Group	
Chief		Division / Group	
Deputy		Division / Group	
Resources Unit		d. Air Operations Branch	
Situation Unit		Branch Director	
Documentation Unit		Air Attack Supervisor	
Demobilization Unit		Air Support Supervisor	
Technical Specialists		Helicopter Coordinator	
Human Resources		Air Tanker Coordinator	
Training		10. Finance Section	
		Chief	
		Deputy	
		Time Unit	
		Procurement Unit	
8. Logistics Section		Compensation/Claims	
Chief		Cost Unit	
Deputy			
Supply Unit			
Facilities Unit		Prepared by (Resource Unit Leader)	
Ground Support Unit			
Communications Unit			
Medical Unit			
Security Unit			
Food Unit			

ICS 203

COMMUNICATIONS PLAN

1. Incident Name		2. Date	3. Time	4. Operational Period

5. Basic Communications Utilization

Assignment	Radio Channel	Function	Frequency	Landline Number	Cell Number	Remarks

6. Prepared by (Communications Unit)

ICS 205

DIVISION ASSIGNMENT LIST		1. Incident Name			
2. Branch	3. Division / Group	4. Operational Period		5. Date	6. Time

7. Operations Personnel			
Operations Section Chief		Division/Group Supervisor	
Branch Director		Air Attack Supervisor	

8. Resources Assigned this Period					
Strike Team/Task Force/ Resource Designator	Leader	Number Persons	Trans. Needed	Drop Off Point/Time	Pick Up Point/Time

9. Control Operations

10. Special Instructions

9. Division/Group Communication Summary							
Function	Frequency	System	Channel	Function	Frequency	System	Channel
Command				Logistics			
Tactical Div/Group				Air to Ground			
Prepared by (Resource Unit Leader)			Approved by (Planning Section Chief)			Date	Time

ICS 204

MEDICAL PLAN	1. Incident Name	2. Date	3. Time	4. Operational Period

5. Incident Medical Aid Station

Medical Aid Stations	Location	Paramedics	
		Yes	No

6. First Response Agency

Agency	Location	Paramedics	
		Yes	No

7. Transportation
A. Ambulance Services

Medical Aid Stations	Location	Paramedics	
		Yes	No

B. Incident Ambulances

Medical Aid Stations	Location	Paramedics	
		Yes	No

8. Hospitals

Name	Address	Travel Time		Phone	Helipad		Burn Center	
		Air	Ground		Yes	No	Yes	No

9. Medical Emergency Procedures

Prepared by (Medical Unit Leader)	Reviewed by (Safety Officer)

ICS 206

Index

145